**OC**

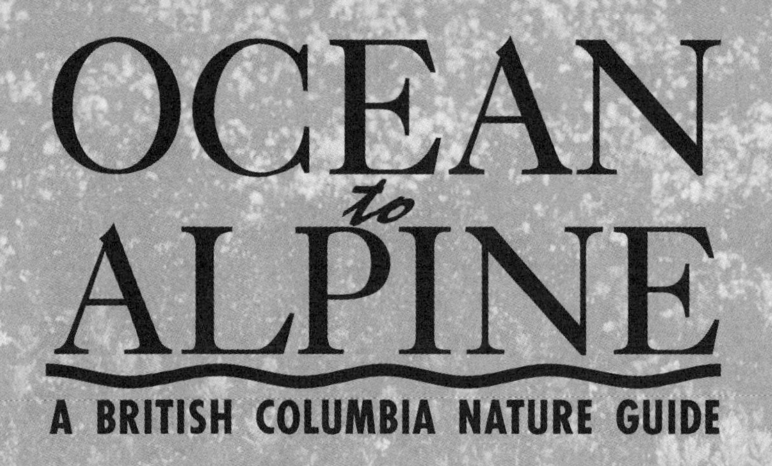


A BRITISH COLUMBIA NATURE GUIDE

Editors

JOY & CAM FINLAY

Printed in Canada

First printed in 1992 5 4 3 2 1



*The publisher:*
Lone Pine Publishing
206, 10426-81 Avenue
Edmonton, Alberta, Canada
T6E 1X5

**Canadian Cataloguing in Publication Data**

Main entry under title:

Ocean to alpine

Includes index.
ISBN 1-55105-013-7

1. Wildlife watching—British Columbia—Guidebooks. 2. Natural history—British Columbia—Guidebooks. 3. British Columbia—Guidebooks. I. Finlay, Joy. II. Finlay, Cam.
QI60.034 1992 574.9711 C92-091400-4

Cover design: *Beata Kurpinski*
Editorial: *Elaine Butler*
Design and layout: *Beata Kurpinski, Tanya Stewart, Bruce Keith*
Maps: *Inge Wilson*
Printing: *Kromar Printing Limited, Winnipeg, Canada*

Front Cover Photo: *C. Pettitt*
Back Cover Photos: *Jim Butler*: Orca, Indian paintbrush
*Canadian Parks Service:* Firetower at Mt Revelstoke National Park

**Photo Credits:** *B.C. Provincial Parks:* 66 (top), 73 (top), 183 (top); *Butler, J.:* 72 (top); *Courtesy of the Royal British Columbia Museum, Victoria, B.C.:* 66 (bottom), 68 (top), 69 (top), 69 (bottom); *Fallis, Mary:* 72 (bottom), 74 (top), 75 (top), 76 (bottom), 77 (bottom), 78 (bottom), 79 (bottom), 80, 179 (bottom), 180 (bottom), 182, 184 (bottom), 185, 187 (top), 188 (top), 191 (top), 192: *Grass, Al:* 67 (top), 74 (inset), 75 (bottom), 77 (top), 79 (top), 178 (bottom), 181 (bottom), 186 (bottom), 190 (top), 190 (bottom), 191 (bottom); *Gyug, Les, Canadian Parks Service*: 178 (top); *Lynch, W., Canadian Parks Service*: 188 (inset); *McCrory, D.*: 76 (top); *McCrory W.*: 186 (top); *Pettitt, C.:* 78 (top); *Sutherland, Bob*: 65, 67 (bottom), 68 (bottom), 70, 73 (bottom), 179 (top); *Tourism B.C.:* 74 (bottom); *Woods, John, Canadian Parks Service:* 177, 180 (top), 181 (top), 183 (bottom), 184 (top), 187 (bottom).

The publisher gratefully acknowledges the assistance of the Federal Department of Communications, Alberta Culture and Multiculturalism, and the Alberta Foundation for the Arts in the production of this book

# CONTENTS



## VANCOUVER ISLAND



## SOUTHWEST BRITISH COLUMBIA

## *SOUTH AND SOUTHEAST BRITISH COLUMBIA;*

## *CENTRAL BRITISH COLUMBIA*



## *NORTHERN BRITISH COLUMBIA*

# PREFACE

This book is a guide for people who enjoy the out-of-doors, whether their interest is hiking, fishing, canoeing, bird-watching, sitting on a knoll looking at a crimson paintbrush, or eating lunch under a tree. It is for those who want to get off the beaten path into our wildlands.

Fortunately B.C. offers plenty of opportunity to do just that. There are sites set aside for every reason and of every size. They range from a grove of Douglas-fir with scolding squirrels, a seashore with harbour seals, alpine peaks with meadows bursting with flowers and small groups of caribou, sheep and goats, or a grassland slough full of mallards to an urban park with singing birds. It has been a labour of love to research and search out so many places. The numerous people across the province who supplied information on special sites and places made this book possible.

We gratefully acknowledge these contributions which were the invaluable data base upon which this book was built. We also thank ever so many others in British Columbia, including those in the Royal British Columbia Museum, and other government departments, Provincial and National Parks.

We thank you ...

A special thanks to Yorke Edwards, former director of the Museum, who first encouraged us to take on this gigantic task; to Jude Grass, who was then president of the Federation of B.C. Naturalists, and who piloted the concept through the many channels of naturalists across the province ensuring we had the utmost cooperation most necessary for such a book. Staff at the Museum have continued to lend their support throughout the project. No book can be finished without the helping hand of a publisher. We thank the people at Lone Pine including publisher Grant Kennedy, editor-in-chief Gary Whyte, and especially Elaine Butler who directly oversaw the work, plus the other staff who were so very positive and helpful throughout the process.

Now to you who use this book, make notes on what you see and send comments to us. We hope to revise this book in about five years so we need your assistance to ensure it is up to date and useful.

Joy and Cam Finlay — 1992

Box 8644, Station L

Edmonton, Alberta

T6C 4J4

## *HOW TO USE THE BOOK: AN OUTLINE*

The information in this book covers the province with a greater emphasis on the more heavily populated regions of the south and southwest. The text starts with a brief lead in sentence or two giving highlights of the site including birds, mammals, amphibians, reptiles, fish, insects and special plants, and covers:

**Birds**: anything special and rarities, if any, such as trumpeter swans or peregrine falcons.

**Mammals** or **Marine mammals**: particularly the large and distinctive ones.

**Reptiles and Amphibians**: since they are hard to locate, only a brief mention, if any, is made. For more help ask local people.

**Fish** or **Marine life**: only briefly, if at all.

**Insects**: including butterflies, on which we have very little information. We need your help.

**Plants** (including trees): which include mainly the flowering plants. Since there are so many species, we have given only the special ones on which we have information.

**Geology:** a short section, but important. This is the base on which everything else depends.

And sometimes special spots to visit and trails to hike.

The **Further Information** section covers how to find the site. Some special hiking trails may also be listed here as well as further references.

# BRITISH COLUMBIA REGIONS

# SPECIAL ATTRIBUTES

## BIRDS

British Columbia, with its tremendous diversity of climates and habitats, boasts more bird species than any other Canadian province; a recent Provincial Museum checklist includes 439 species, of which 292 are known to nest here. Although a number of bird species (e.g., hairy woodpecker, common raven, American robin, pine siskin) are found throughout British Columbia, most of them are confined to particular regions. The islands and near-shore waters of the Pacific coast, the verdant coastal coniferous forests, the arid valleys of the southern interior, and the boreal forests and parklands of northeastern B.C., each have many unique species, and contribute importantly to the overall variety of bird life.

Peak migration times range from March 1 to May 31, and August 10 to November 20. These are the most interesting times to watch birds in B.C. The variety of species is greatest, and new species arrive while others depart every week. Movements of small landbirds are often unspectacular by eastern Canadian standards, partly because the migration seasons, at least in southern B.C., are quite extended rather than being compressed into a few short weeks. However, some huge "waves" of warblers and other migrants do occur, especially in spring from late April to early May. Movements of waterfowl and other water-birds in some areas are impressive, such as the migration of snow geese, brant, and ducks on the south coast, Canada geese and sandhill cranes in the Okanagan Valley, and tundra swans and ducks in the Creston area.

Offshore bird movements are often remarkable. Huge flocks of sooty shearwater pass by in late summer, and scoters, phalaropes, jaegers and Sabine's gull move through in the spring and fall. However, perhaps the most spectacular migratory concentrations in B.C. are those of the shorebirds. There are several major stopping places. The largest, by far, is the Fraser River estuary, including Boundary Bay. Western sandpiper and dunlin dominate the scene, with both species occurring in the tens of thousands. Others, including black-bellied plover, sanderling, least and pectoral sandpipers, and short-billed and long-billed dowitchers, also appear in the hundreds. A few "casual" or "accidental" species are found every year including the spotted redshank, far eastern curlew, bar-tailed godwit, curlew sandpiper, and spoonbill sandpiper.

The variety of wintering birds increases as you move from the north and central interior of the province to the south interior, with the largest numbers on the west coast. Nearly 50 Christmas Bird Counts are now carried out in B.C. each year between December 15 and January 4, and counts at Ladner, Vancouver and Victoria consistently report 130 species or more — the highest diversity of winter bird life in the nation. In 1987 Victoria set the Canadian record with 145.

Wintering birds in northern and central B.C. are generally limited to a few hawks and owls, woodpeckers, corvids, chickadees, nuthatches and finches. Southern valleys have a much larger number of permanent resident and wintering landbird species, and also have large lakes and rivers which rarely freeze over, affording habitat for a variety of waterbirds. Notable winter concentrations of waterbirds in the southern interior include hundreds of tundra swans on the South Thompson River and thousands of American coots and redheads on Okanagan Lake. However, by far the largest wintering concentrations of waterbirds occurs on the coast. The Fraser estuary in particular, and smaller estuaries or shallow bays elsewhere, support thousands of mallard, American wigeon, northern pintail, green-winged teal, greater scaup, and dunlin. Every cove or small inlet seems to have its flock of surf scoter and Barrow's goldeneye, and huge rafts of western grebe winter in a few favoured localities. Although large flocks of waterbirds are localized, few stretches of shoreline are bereft of birds. This abundance of coast wintering birds stands in contrast to summer, when many of the shallow coastal waters are almost birdless, unless there is a nearby seabird colony.

*Rock Wren*

## BIOTIC AREAS

Biotic areas as described by Munro and Cowan in their book *A Review of the Bird Fauna of British Columbia* (1947) are a convenient framework for describing the distribution of breeding birds in the province.

### COAST LITTORAL BIOTIC AREA

This zone includes thousands of small islands along the B.C. coast, as well as the coastal waters. Many of the islands support nesting sea-bird colonies. In the protected waters of the Gulf of Georgia, these colonies include glaucous-winged gull, and double-crested and pelagic cormorants. Around the Queen Charlotte Islands and along the west coast of Vancouver Island, sea-bird colonies support many additional species, including fork-tailed and Leach's storm-petrels, Brandt's cormorant, common and thick-billed murre, ancient murrelet, Cassin's and rhinoceros auklets, and tufted and horned puffins. Although many of these islands are difficult to access and require a special permit to visit, a trip to one of them is an experience never to be forgotten.

### PELAGIC WATERS BIOTIC AREA

This zone, consisting of oceanic waters more than 15 km or so offshore, is home to a group of bird species which are rarely or never seen from land. For the average naturalist, the easiest way to visit this zone is to reserve a space on a pelagic birding trip from Bamfield, Ucluelet, or Tofino on the west coast of Vancouver Island. Birds to be expected include black-footed albatross, northern fulmar, shearwaters (pink-footed, flesh-footed, Buller's, and sooty), phalaropes, jaegers (pomarine, parasitic and long-tailed), south polar skua, black-legged kittiwake, and Sabine's gull, plus the storm-petrels and alcids named in the preceding paragraph.

COAST FOREST BIOTIC AREA

These lands include most of the western slopes of the coastal mountain ranges. Typical breeding birds of this humid hemlock and cedar forest include blue grouse, marbled murrelet, band-tailed pigeon, red-breasted sapsucker, western flycatcher, Steller's jay, chestnut-backed chickadee, winter wren, golden-crowned kinglet, and varied thrush. None of them is confined to this biotic area, but most are commoner here than elsewhere.

QUEEN CHARLOTTE ISLANDS BIOTIC AREA

This area is similar to the Coast Forest in its birds, but lacks several species common on the mainland. Several subspecies are found here including hairy woodpecker, Steller's jay and song sparrow.

GULF ISLANDS AND PUGET SOUND LOWLANDS BIOTIC AREAS

These two areas are treated as one unit for this book because the birds found here are very similar although the mammals are quite different. Drier than the Coast Forest, these zones were originally dominated by Douglas-fir, arbutus, and Garry oak, but are now largely cropland and pasture. Typical forest and forest-edge species include the bushtit, Bewick's wren, Hutton's vireo, black-throated gray warbler (mainland only), and western tanager. The larger rivers and their bordering cottonwoods are home to the green-backed heron, wood duck, and black-capped chickadee (mainland only). Introduced bird species are prominent, including the crested myna around Vancouver and common skylark on Vancouver Island.

DRY FOREST BIOTIC AREA

This zone, including the extensive ponderosa pine forests and adjacent grasslands and sagelands of the southern interior, supports a whole complex of bird species shared only, if at all, with the Osoyoos Arid Area. Typical bird species include poor-will, calliope hummingbird, western kingbird, mountain chickadee, white-breasted and pygmy nuthatches, rock wren, western bluebird, gray catbird, Nashville and yellow-rumped warblers, western tanager, lazuli bunting, vesper sparrow, western meadowlark, and Cassin's finch.

OSOYOOS ARID BIOTIC AREA

Occupying the lower Okanagan and Similkameen Valleys, this area is mainly orchards and vineyards. In places there are numerous cliffs and rocky outcrops which harbour many of the distinctive bird species, including prairie falcon, chukar (introduced), white-throated swift, canyon wren, sage thrasher, Brewer's sparrow. Along streams look for American redstart and yellow-breasted chat, plus many of the Dry Forest species.

## COLUMBIA FOREST BIOTIC AREA

Occupying many valleys in the Kootenay region, this humid low-altitude forest area is somewhat like the Coast Forest. The birds include a selection of those present in the Coast Forest, Dry forest, and Subalpine Forest areas. Although no species is restricted to the Columbia Forest, barred owl is more numerous here than elsewhere. Other common birds include Hammond's flycatcher, chestnut-backed chickadee, red-breasted nuthatch, brown creeper, golden-crowned kinglet, yellow-rumped warbler, cedar waxwing, black-headed grosbeak, and Cassin's finch.

## CARIBOO PARKLANDS BIOTIC AREA

Cariboo Parklands is somewhat transitional between the Dry Forest and Subalpine Forest, including extensive Douglas-fir forests, open grasslands, and many aspen groves. The Cariboo's unique feature is the many small wetlands, home to common loon, white pelican (on Stum and Alkali lakes), American coot, black tern, and many duck species; this is B.C.'s "duck factory." Western meadowlark, vesper sparrow and horned lark inhabit the grasslands, while yellow-bellied sapsucker, western wood pewee, tree swallow, and mountain bluebird nest in nearby aspen groves.

## SUBALPINE FOREST BIOTIC AREA

Subalpine forest area, dominated by Englemann spruce and subalpine fir forests, occurs from northern B.C. to the American border. Birds include northern goshawk, spruce grouse, three-toed woodpecker, olive-sided flycatcher, gray jay, Clark's nutcracker, boreal and mountain chickadee, ruby-crowned kinglet, hermit and varied thrushes, Townsend's and Wilson's warblers, fox and Lincoln's sparrows, pine grosbeak, and white-winged crossbill.

## PEACE RIVER PARKLANDS BIOTIC AREA

This outlier of aspen parkland of the prairie provinces, consists of aspen forest, now largely replaced by grain fields, with a few patches of open grassland. Conifers are mainly found in the major river valleys. The region, lying east of the Rocky Mountains, has bird life that is overwhelmingly eastern, and includes many species not found elsewhere in B.C. Typical birds include sharp-tailed grouse, alder and least flycatchers, eastern phoebe, blue jay, black-billed magpie, house wren, Philadelphia vireo, and many "eastern" warblers, including Tennessee, black-throated green, black-and-white, American redstart, ovenbird, Connecticut, mourning and Canada warblers. Other species include rose-breasted grosbeak, clay-coloured sparrow, white-throated sparrow, and northern oriole. The scattered wetlands are home to eared grebe, black tern, Le Conte's and sharp-tailed sparrows, and common grackle.

BOREAL FOREST BIOTIC AREA

These lands are covered mainly with white and black spruce forest and some deciduous aspen, birch and willow. The area occupies most of northeastern B.C., plus most major river valleys in the northern interior. Wetlands are numerous in the Fort Nelson Lowland region. Many bird species are shared with the Peace River Parklands and Alpine Forest. Characteristic birds include northern goshawk, spruce grouse, northern hawk-owl, great gray owl, alder and yellow-bellied flycatcher, gray jay, black-capped chickadee, Tennessee, magnolia, yellow-rumped, palm, bay-breasted, and blackpoll warblers, swamp sparrow, and rusty blackbird.

NORTHERN ALPLANDS AND SOUTHERN ALPLANDS BIOTIC AREAS

Bird life is not very numerous in these areas. Only five species — white-tailed ptarmigan, rock ptarmigan (found only in northern B.C.), horned lark, water pipit and rosy finch — regularly nest above timberline. The stunted tree area at, and just below, timberline provides habitat for several other species including fox and golden-crowned sparrows. In addition, in northern B.C. look for willow ptarmigan, gray-cheeked thrush, American tree sparrow, and Smith's longspur.

## ***FURTHER INFORMATION***

Many pamphlets, checklist and books have been written on the birds of B.C. including the recent release of one by the Cannings (*Birds of the Okanagan Valley, British Columbia*) that should become a classic for regional studies. Some of these include:

Beebe, F.L., 1974. *Field Studies of the Falconiformes of British Columbia.* British Columbia Provincial Museum, Occasional Paper 17. 163 pages.

Campbell, R.W., H.R. Carter, C.D. Shepard and C.G. Guiguet, 1979. *A Bibliography of British Columbia Ornithology*, Vol. 1, and Vol. 2. British Columbia Provincial Museum, Heritage Record Series No. 7.

Cannings, R.A., R.J. Cannings, S.G. Cannings, 1987. *Birds of the Okanagan Valley, British Columbia.* Royal British Columbia Museum. 420 pages.

Guiguet, C.J. 1954 to 1983. *The Birds of British Columbia.* British Columbia Provincial Museum Handbooks No. 6, 8, 10, 13, 15, 18, 22, 29, 37, and 42.

Holroyd, Geoff and Howard Coneybeare. 1989. *Compact Guide to Birds of the Rockies.* Lone Pine Publishing, Edmonton, Alberta.

Mark, D.M. 1984. *Where to Find Birds in British Columbia.* Second edition. Kestrel Press, New Westminster, B.C. 122 pages.

Munro, J.A. and I.McT. Cowan. 1947. *A Review of the Bird Fauna of British Columbia*. British Columbia Provincial Museum, Special Publ. No 2. 285 pages.

Weber, W.C. 1980. "A Proposed List of Rare and Endangered Bird Species for British Columbia." Pages 160-182 in R. Stace-Smith, L. Johns, and P. Joslin (eds.), *Threatened and Endangered Species and Habitats in British Columbia and the Yukon*. British Columbia ministry of Environment, Victoria. 302 pages.

*Wayne Weber*

## MAMMALS

British Columbia probably has the most diverse combination of climatic and geological features anywhere in North America. This diversity is reflected in the wide variety of mammals that are found in the province. Of the 196 species of mammals found on land or in adjacent waters in Canada, B.C. has 139. Moreover, the province's population of several of these species, including the California bighorn sheep, Stone sheep, mountain goat and cougar, is probably the largest in the world.

Many of the land mammals that are exclusive to the province are relatively common and provide excellent opportunities for the viewing public. The variety and abundance of large land mammals is what sets B.C. apart from most other wildlife areas across North America. Big mammals include moose, elk (wapiti), caribou, black-tailed and white-tailed deer, bighorn and Dall sheep, bison, mountain goat, grizzly and black bears, cougar, wolf, fox, coyote, lynx and wolverine.

Many of these large animals may be viewed relatively easily during the summer months. A trip along several highways through the mountains could result in sightings of two species of deer, elk, moose, black and grizzly bears, mountain goat and mountain sheep. Even more spectacular views are available in winter when some of these big animals gather on winter ranges near valley floors and close to highways. Although travel by motor vehicle can be hazardous in winter, some of the best viewing opportunities for both wildlife, especially ungulates, and scenery occur at that time of the year.

Most carnivores are hard to locate because of their scattered distribution, their often solitary life style and secretiveness. Tracks are usually the only indication they have been around. Grizzly bears, however, do congregate along some streams and rivers when the major salmon runs take place in the fall. These magnificent creatures sometimes provide incredible viewing opportunities, but no one should embark on such expeditions without a thorough knowledge of these mammals and the potential danger they can be. Ensure

*Pika*

you have someone to accompany you who is thoroughly familiar with the behaviour of bears. If the proper care is taken, they can be observed in relative safety (see *Bears* sidebar).

Coyotes are widely distributed throughout the province and generally prefer open country, which makes them easier to see. Many happy hours can be spent watching a coyote search for mice and other small rodents, which form a major part of its food.

The cougar, also known as the mountain lion or puma, occurs wherever a suitable source of food, principally deer, may be found. The sighting of a cougar is likely to be the highlight of any trip. Look for tracks along soft spots on a trail, or edges of streams where deer come to water. Winter is an excellent time to find cougar tracks in areas of heavy deer concentration.

***BEARS: BE INFORMED AND TRAVEL SAFELY***

*Many of B.C.'s parks and wilderness areas are inhabited by black bears and grizzlies. Visitors to these areas should be well informed and take proper precautions when traveling in bear country, not only for your safety but to preserve the bears as well.*

*Grizzlies are more frequently found in the back-country and range from the rugged coastal mountains to the Great Divide of the Rockies. Although bears may be encountered almost anywhere, learning to identify key bear habitats/feeding areas, such as avalanche paths, will minimize your chances of surprising one. In general, keep together as a group and make loud noises to warn bears of your presence, particularly along winding trails with dense vegetation where it is difficult to see ahead and along noisy creeks.*

*Learn about proper storage of your food and garbage before hiking or camping, to avoid attracting bears.*

*While chances of a serious encounter are rare (it is usually more dangerous to drive on the highway than to hike in bear country), it is important to leave your wilderness itinerary with a park warden, ranger, forest manager, R.C.M.P. or friend in case of emergency. Do so even on a day trip.*

*It is a thrilling experience to see a bear, but when you do, respect it for what it is -- a wild, potentially dangerous animal, not to be approached or provoked.*

*The most recommended guide is Bear Attacks: Their Causes and Avoidances by Stephen Herrero.*

*Erica Mallam and Wayne McCrory*

Marine mammals are well represented in the province by whales, porpoises, seals and sea lions. Killer whales, seals and sea lions may be regularly seen along the coast, often while travelling on the ferry system between Vancouver and

Vancouver Island. Sea lions are especially common in winter, and gray whales regularly migrate along the west coast of Vancouver Island in the spring.

Some of the smaller mammals are often easy to locate and watch. Pika occur on talus slopes in high mountain regions and build their own haystacks for a winter food supply.

Squirrels and chipmunks are familiar to nearly everyone and provide hours of enjoyment. These small mammals are most everywhere in the province, particularly in the forest areas. Both the red squirrel and northern flying squirrel are present, together with several species of ground squirrels and marmots. Another rodent found in the province is the mountain beaver, which is not a beaver at all and looks more like a muskrat. The true beaver is present in large numbers, especially where there are aspen, its favorite food. High-country hikers will soon find the hoary marmot, largest of the marmots, weighing up to 15 kg. Bushy-tailed wood rats, otherwise known as pack rats for their habit of stealing objects from a camp site, are also a familiar mountain species.

Many species of shrews and moles, members of the order *Insectivora*, are native to B.C. The water shrew spends most of its time feeding on underwater insects and may surprise a swimmer or fisherman. Moles only occur in the Fraser valley. The bat fauna of B.C. is very diverse and easy to see under campground and rural street lighting. B.C. also has one marsupial — the opossum, which expanded into the province in 1949 from a population introduced in Washington State. Today this cat-sized mammal is very common throughout the lower Fraser Valley as far as Spuzzum, but it has not yet gone north of the Fraser River.

### *FURTHER INFORMATION*

Few general mammal references are readily available for the province. We continue to use *The Mammals of British Columbia* by I. McT. Cowan and C.J. Guiguet, 1965, third edition. B.C. Provincial Museum, Handbook No 11.

Another helpful book is *Mammals of the Northwest, Washington, Oregon, Idaho and British Columbia* by E.J. Larrison. 1976. Seattle Audubon Society.

*W. T. Munro and Chris Dodd, both of the B.C. Wildlife Branch*

## Seashore Life

To get the most out of a trip to the seashore, try to coordinate your visit with a low tide of two feet or less. Most local newspapers print the daily tides for the area.

### EXPOSED ROCKY SHORE

This type of shore typifies the west coast of Vancouver Island and the Queen Charlotte Islands. Here the open Pacific swells pound into the jagged rocks of the coast. The animals and plants need to be tough and tenacious to withstand the pounding and sucking waves.

Beds of California mussels dominate this type of coast-line. Dense colonies of these bivalve molluscs cling to the upper rocks with strong bushy threads. They sometimes pile on top of each other 20 cm deep. Mussels are filter-feeders. They suck food-laden water through the gills which filter out the particles and channel them to the mouth. The filtered water then passes out through a gap between the shells. Mussels feed primarily in spring and summer when surface waters teem with a rich soup of plankton, single-celled plants and minute shrimplike animals.

The barnacle, another common filter feeder, cements itself to the rocks just above the mussels. They are related to crabs, although from their outward appearance this is hard to imagine. Inside the volcano-shaped shell, a soft-bodied animal lies on its back with its hairy, jointed legs sticking up through an opening in the top. In this position, the animal waves its hairy limbs through the water picking up small food particles from the plankton. It then retracts the limbs and scrapes the food off. There are three common kinds of acorn barnacles in the intertidal zone, each living at a different tidal level. The goose barnacle, although similar internally, appears unrelated. Instead of being cemented directly to the rock, the shelled end attaches to a rubbery stalk that bends and flexes with the rushing waters. Protruding its filtering legs out into the wave surge, it lets the water do the work.

Four common species of limpets, or "Chinese Hats," occur between the high and low tide. Each prefers a different amount of air exposure. The small finger limpet can survive prolonged exposure to air so it lives highest on the shore.

In the upper shore, tiny (5 to 10 mm) black snails called periwinkles abound. In the splash zone, wetted occasionally by spray, a relative of the wood bug, the rock slater, nestles in the cracks or skitters across the bare rocks.

At the top of the mussel bed, a seaweed called bladderwrack establishes itself. This seaweed gets its common name from the gas-filled bladders at the end of its branches. Another unique form, the sea palm, grows on mussels but only in areas that receive the direct impact of ocean swells. It is so named because it resembles a tiny palm tree. Brown leathery seaweeds predominate on this exposed shore-line. Thicker, robust species take over closer to low water. Hiding among the labyrinth of spaces within the mussel bed, a myriad of other animals survive in this semi-protected environment — porcelain crabs, limpets, whelks, tarspot sea cucumbers, isopods (marine pill bugs), polychaete worms, ribbon worms, and small aggregating anemones.

Common Ochre nestle in groups on the lower edge of the mussel bed. These sea-stars, with their patterns of white markings on a purple or yellow background, feed on a mussel by surrounding it with their protruded stomach. While the stomach juices slowly digest the soft parts, rows of suckers exert a slow steady pressure to pull the shells apart. Purple stars cannot survive out of water for long so they only venture half way up the beach. As a result, they crop the mussels up to that level, leaving a distinct lower edge to the bed. If these predators were removed, mussels would probably cover most of the intertidal zone!

Other species here include the bright red blood star, a small six-legged star, the thatched barnacle, and the black chiton. In more sheltered areas the black turban snail buries itself in the gravel around the bases of boulders.

Closer to the low-tide level, lush brown seaweeds flourish, with names like feather boa, sea sac, sea cabbage, spongy cushion, triple rib and sea girdle. What appear to be splashes of pink paint on the rock are actually a plant. Other forms of this calcareous alga grow in tufts around the rims of tidal pools. Their jointed branches, when bleached white, are often mistaken for coral.

In the lower intertidal, emerald green surf grass adorns the rocks. But this is no ordinary seaweed. It is a marine seed plant with microscopic flowers that are pollinated underwater. Most other seaweeds seen on the shore are algae. They have chlorophyll but their "roots" are simply anchors. Nutrients are absorbed directly from the water, and reproduction is by spores.

Green sea anemones and purple sea urchins create splashes of colour in tide pools, at the base of rocky slopes and in surge channels.

Many more plants and animals dwell in this exposed, intertidal zone than can be described here. The list of books at the end of this section can provide more information about this amazingly rich part of the coast.

## EXPOSED SAND

Compared to rocky exposed coast, sand beaches are relatively sparse in marine life. However, some species live only on these beaches. Razor clams, for instance, can quickly bury their thin, elongate shells beneath the surface of this hard-packed sand. In softer sand, the bullet-shaped, pinkish-grey olive shell burrows just below the surface, leaving a well-marked trail behind it. Higher up this soft sand beach a section of pockmarks indicates the presence of either thin red blood worms or swarms of sand fleas. The latter emerge at night to feed on the dead seaweed stranded at the high tide line.

Many treasures await the curious beachcomber: tangled masses of bull kelp (a subtidal alga renowned as one of the fastest-growing plants in the world); driftwood festooned with clusters of leathery, stalked goose barnacles, relatives of the rocky-shore kind but with thinner shells and smoother stalks; purple sailor jellyfish driven ashore by storm winds to form windrows of dried bodies. Shaped like an oval plate turned upside down with a stiff diagonal sail across the top, these open-ocean creatures drift with tides and winds. On the underside, hundreds of purple polyps make up the rest of the colony.

## PROTECTED ROCKY SHORE

Here, among protected islands and sheltered inlets, waves are small and freshwater from rivers and creeks dilutes the sea. In this environment similar animals and plants, like mussels, snails, chitons, and crabs exist, but many are different species from those on the exposed coast. For example, the blue mussel replaces the more robust California mussel because it can withstand the dilute sea water. The common wrinkled whelk lays eggs in tiny vaselike capsules which it attaches to the underside of rocks in mid winter. Inside the capsule the yolky eggs develop into tiny snails in about 2 or 3 months. When they emerge they are ravenous! Many baby barnacles fall prey to these tiny predators.

In quiet waters the purple star is less common. The flatter mottled star replaces it. On a low tide the smooth-skinned leather star, with its wide-based arms, may be hiding under the seaweeds. Orange sea cucumbers, red rock crabs with black-tipped claws, snails like the dire whelk and leafy hornmouth, long rubbery ribbon worms, tubeworms, pillbugs and little neck clams all compete for space and food at this lower level.

PROTECTED SAND

Many popular beaches on the east coast of Vancouver Island fall into this group. Thousands of sand dollars — flattened relatives of the sea urchin — live in clean sand near the lowest tide. If you are lucky, you may find a grey object resembling the end of a toilet plunger. This collar made of compacted sand contains the eggs of a moon snail. When fully extended, the soft foot of a moon snail may be three times larger than the 8 cm shell and completely surround it. An empty clam shell with a neat countersunk hole illustrates the work of this predatory snail.

Where there are moon snails there are clams, like butter clams, horse clams, and soft shell clams; but **beware of "red tide" or Paralytic Shellfish Poisoning.** When a massive bloom of toxic single-celled plankton is filtered by a clam, the toxins accumulate in the clam's tissue. To avoid over-harvesting, be aware of limits set and enforced by the Fisheries Department.

In protected bays where the sand is softer, you will see several kinds of holes. Those surrounded by a small mound of sand are probably made by shrimp; clams produce a hole with no mound; and lugworms create a small mound topped with a coil of sand. Extensive beds of eelgrass flourish in these quiet bays. Like surf grass, this seaweed is a vascular plant adapted to life in the ocean. All sorts of animals live on and under the fronds of eelgrass, including bubble snails, sea anemones, green isopods, sea slugs, Dungeness crabs, shrimp, shiner perch, staghorn sculpins and sand dabs.

***FURTHER INFORMATION***

B.C. Provincial Museum Handbook Series:

Quayle, D.B. *Intertidal Bivalves of B.C.* No. 17.

Griffith, L.M. *Intertidal Univalves of B.C.* No. 26.

Lambert, P. *Sea Stars of B.C.* No. 39.

Hart, J.F.L. *Crabs and their Relatives of B.C.* No. 40.

Carefoot, T. 1977. *Pacific Seashores: A Guide to Intertidal Ecology.* J.J. Douglas Ltd. Vancouver.

Guberlet, M.L. 1974. *Seaweeds at Ebb Tide*. University of Washington Press, Seattle.

Hewlett, S. and K. G. Hewlett. 1976. *Sea life of the Pacific Northwest*. McGraw-Hill Ryerson. The plates are a mixture of colour and black and white.

Johnson, M. E. and H. J. Snook. 1927. *Seashore Animals of the Pacific Coast*. Dover Edition 1967. Dover Publications Inc. New York. This is an old publication but still very useful.

Kozloff, E. N. 1983. *Seashore Life of the Northern Pacific Coast*. Douglas and McIntyre, Vancouver.

Rickets, E. F. and J. Calvin. 1968. 4th Ed. *Between Pacific Tides*. Stanford University Press, California.

Smith, Lynwood S. 1976. *Living Shores of the Pacific Northwest*. Pacific Search.

*Philip Lambert*

## Insects

The province encompasses many habitats, from the wettest to the driest in Canada, and from sea level to some of the highest mountains in the country. Because of this diversity probably 30,000 insect species occur in the province. Our incomplete knowledge of them is emphasized by the fact that only about one-half are known and described. Only a handful of the more common or interesting species are mentioned here.

Our present insect fauna is relatively young, having recolonized the region after the glaciers melted about 10,000 to 15,000 years ago. Most species survived the glacial period south of the ice, but large, ice-free areas in the Yukon and Alaska apparently harboured species while ice covered the rest of the northern landscape. There is some evidence of smaller ice-free "refugia" on the outer west coast, especially on the Queen Charlotte Islands and northern Vancouver Island.

The coastal region is characterized by wet forest. The insect orders *Coleoptera* (beetles) and *Diptera* (true flies) dominate these forests because a vast array of their species develop in decaying wood, leaf litter and fungi. Ground beetles (*Carabidae*) and rove beetles (*Staphylinidae*) are especially noticeable. Large ground beetles called *Scaphinotus* feed on the slugs so abundant in these forests. A particular rove beetle, *Pelecomalium testaceum*, is common on the yellow scapes of the skunk cabbage in spring. These beetle families are also prevalent on the long, sandy beaches along the outer coast. At night, a naturalist with a flashlight can spot many kinds out hunting over the sand. The large, wingless *Thinopinus pictus*, banded yellow and brown, is especially striking. Clouds of various primitive, slender-legged flies teem in the nearby dark forest; these include fungus gnats, midges, crane flies, and the unique *Cramptonomyia spenceri*. The latter is a species whose larva tunnels in red alder logs, and is one of only four species of *Pachyneuridae* in the world, and the only one in North America. The marsh crane fly, *Tipula palustris*, is an immigrant from Europe whose larvae, called leather jackets, periodically devastate lawns by feeding on the grass roots, in the south coastal cities. The large adult flies with their long gangly legs are particularly numerous in September, when they often enter houses.

In the drier Douglas-fir forests of the coast, long-horned (*Cerambycidae*) and metallic wood-boring (*Buprestidae*) beetles are more evident. Cerambycids range from the huge pine sawyer (*Ergates spicularis*), up to 7 cm long and the laurel borer (*Rosalia funebris*), with its elegant black and grey banding and 5 cm long antennae, to the strange, bee-mimicking lion beetle (*Ulochaetes leoninus*). In the Buprestidae, the iridescent green golden buprestis (*Buprestis aurulenta*) is an attention-getter.

Some forest insects are notable because of the damage they do to wood. Some also attack the wood in our homes, and are most conspicuous when they swarm to mate -- carpenter ants (*Camponotus herculeanus*) in May, and termites (*Zootermopsis angusticollis*) in late August. Only termites actually eat the wood; the ants simply build galleries to live in.

A few insects inhabit the marine environment, at least in the intertidal zone. Especially prominent are the intertidal chironomid midges such as *Paraclunio alaskensis* and *Saunderia clavicornis*; the larvae eat algae and diatoms. Several species are introduced members of the seashore fauna -- the cosmopolitan earwig (*Anisolabis maritima*) and the tide-pool mosquito (*Aedes togoi*) apparently have arrived from Asia.

Coastal streams produce an abundance of stoneflies (*Plecopera*), mayflies (*Ephemeroptera*), and caddisflies (*Trichoptera*), which serve as food for fish and birds such as the American dipper. Characteristic dragonflies of these streams are the huge yellow and black *Cordulegaster dorsalis* and the rare *Octogomphus specularis*. Marshes bordering lakes and ponds contain many more species including the striking red dragonfly of spring, *Sympetrum illotum* and the boldly marked *Libellula forensis* and *Libellula lydia* with their brown-banded wings and white abdomens.

Perhaps the most noticeable butterflies are the various swallowtails, including the western tiger swallowtail (*Papilio rutulus*) and the pale swallowtail (*Papilio eurymedon*). The smaller anise swallowtail (*Papilio zelicaon*), whose caterpillar feeds on cow parsnip and other *Umbelliferae*, is probably the most common species on the outer coast. The Lorquin's admiral (*Limenitis lorquini*) and the Sara orangetip (*Anthocharis sara*) are also familiar. The skipper *Erynnis propertius* is found only in the extreme southwest of the province, for its caterpillar feeds only on Garry Oak, a tree restricted to this area. In agricultural areas in parts of southern British Columbia, the cabbage white (*Pieris rapae*) and the common sulphur (*Colias philodice*) may be abundant.

Moths are mostly drab and inconspicuous, but a number are spectacular. The two largest, the polyphemus (*Antheraea polyphemus*) which is light brown with large eye-spots on the hindwings, and the reddish purple ceanothus silkmoth (*Hyalophora euryalis*) are rather common. Their cocoons contain silk. Sphinx, or hawk, moths are varied and beautiful; some fly in daylight and feed at flowers like hummingbirds do.

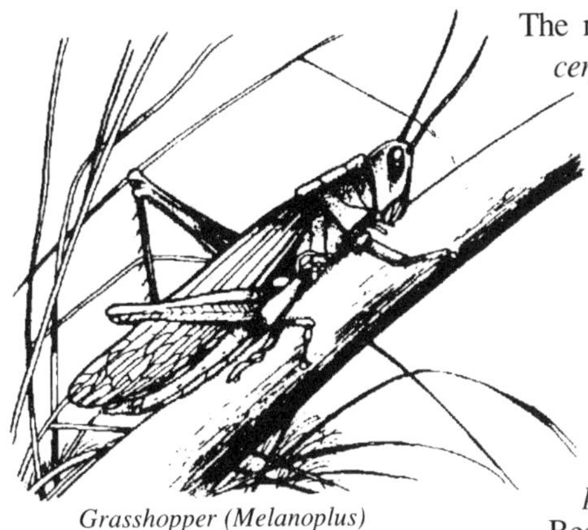

Grasshopper (*Melanoplus*)

The nocturnal eyed hawk moth (*Sphinx cerysi*) is perhaps the most common.

The dry Interior region of the province offers a wide range of insects for study. In the sandy, semi-arid habitats and dry grasslands, large, black darkling beetles (*Tenebrionidae*) trundle across the trails like armored cars. Flighty, sickle-jawed tiger beetles, such as the brilliant green *Cicindela purpurea* buzz from the dusty ground. Beeflies (*Bombyliidae*) hover around sand banks searching for bee and wasp nests to parasitize. Robber flies (*Asilidae*), especially the large orange *Stenopogon inquinatus* or the silvery *Efferia* species, drone after insect prey and capture it in their bristly legs. Solitary wasps of great variety fly about. Velvet ants, a kind of wasp clothed in orange and black hair (with wingless females), scamper across the ground. Colonies of red ants (*Formica obscuripes*) build conspicuous mounds on grasslands or in the dry forests. Wasps of all sorts, including hundreds of species of tiny parasitic ones, abound.

The sounds of *Orthoptera* are everywhere -- the crackle of the yellow-and-black-winged cracker grasshopper (*Trimerotropis verruculata*) and the trilling of the tree crickets (*Oecanthus*). Cicadas (true bugs and not Orthoptera) are shrill in the foliage. The pale, wingless Jerusalem cricket, *Stenopelmatus fuscus*, may be found, although it is nocturnal and more abundant in the pellets of owls than in the collections of sun-loving entomologists! The native ground-inhabiting mantid, *Litaneutria minor*, is rare in the extreme southern Okanagan; the introduced praying mantis (*Mantis religiosa*) is more common. The voracious larvae of ant-lions wait for prey to slide down the walls of the conical holes they dig in sandy soil, and delicate green lacewings with golden eyes crowd around electric lights in summer.

The lakes and streams of the Interior are rich with aquatic insects. These water bodies, such as Okanagan Lake, often produce swarms of short-lived mayflies; the cast larval skins of some species often form windrows on the beaches. Caddisfly larvae, with their curious cases made of sand, pebbles, twigs, or leaves are common in most waters; the moth-like adults flutter around shoreline vegetation. Dragonflies (*Odonata*) are conspicuous. Perhaps most spectacular is the black and white *Libellula pulchella*, which is common around marshes and ponds. Clouds of red *Sympetrum* and blue damselflies of the genus *Enallagma* are characteristic of ponds, even in the waters of salt-ringed alkaline ponds.

The Oregon swallowtail (*Papilio machaon oregonia*) is a characteristic Interior butterfly; the larva feeds on tarragon in the dry country. The two-tailed swallowtail (*Papilio multicaudatus*) is also found in the southern valleys. The common alpine (*Erebia epipsodea*) and the phoebus parnassian (*Parnassius phoebus*) are typical grassland species. Along the streams, anglewings (*Polygonia*) flutter, and a lucky observer may spot a rare monarch (*Danaus plexippus*) among the roadside milkweed. Mud puddles attract hosts of orange and black fritillaries (*Speyeria*), blues (*Lycaeides, Plebejus*), and others.

Lovely underwing moths (*Catocala*) sometimes fly to lights at night, whereas the clearwing hawkmoth (*Hemaris thysbe*), hovers over flowers in the sunlight. Now and then the huge black witch (*Ascalapha odorata*), a rare visitor from Central America, appears. The brown and black banded wooly bear caterpillar of the moth *Isia isabella* is a friendly sight crossing the roads in autumn, looking for a place to overwinter. Some moth caterpillars are less welcome — the fall webworm (*Hyphantria cunea*), forest tent caterpillar (*Malacosoma disstria*), hemlock looper (*Lambdina fiscellaria*, and the Douglas-fir tussock moth (*Orgyia pseudotsugata*) are all serious pests of trees. In the apple orchards the codling moth (*Cydia pomonella*) has long been a major pest.

Some of the most interesting insects are found in the mountains. Insects often "hilltop," or congregate on mountaintops, especially to mate. Subalpine meadows in summer are thick with butterflies -- parnassians (*Parnassius*), arctics (*Oeneis*), checkerspots (*Occydryas*), fritillaries (*Speyeria, Clossiana*), and alpines (*Erebia*). The short-tailed swallowtail (*Papilio indra*) barely makes it into Canada in the Cascade Mountains. In the northern regions the Old World swallowtail (*Papilio machaon*) is common. Also abundant on mountain flowers are hover flies (*Syrphidae*), which often mimic bees and wasps.

Especially evident in butterflies is the increased darkness of wing colour with increasing altitude, which is an adaptation for improving heat absorption at higher, cooler, elevations. This darkening also occurs with a proximity to the cooler Coast, as well as at higher latitudes.

The rare *Tanypteryx hageni*, a primitive dragonfly that in Canada only inhabits the Cascade and southern Coast Mountains, has an amphibious larva that lives in mud burrows in seepage areas. At mountain lakes and northern bogs, the dominant dragonflies belong to the genera *Somatochlora* (metallic blackish-green), *Aeshna* (the blue darners), and *Leucorrhinia* (little red and black species with white faces).

The famous *Grylloblatta* (glacial crawler), a wingless relative of earwigs, is at home only at temperatures around freezing, and preys on other insects in talus slopes and at the edge of glaciers. In winter, skiers may see wingless crane flies (*Chionea*) looking like spiders moving over the snow, *Grylloblatta*, or snow scorpionflies (*Boreus*); in fact *Boreus elegans* is the unofficial provincial insect. Small winter stoneflies (mainly species of *Capnia*) often appear on the snow near streams in early spring when adults emerge from the water through cracks in snow and ice.

### *FURTHER INFORMATION*

Belton, P. 1983. *The Mosquitoes of British Columbia.* B.C. Provincial Museum Handbook No. 41. Victoria. 189 pp.

Cannings, R.A. and K.M. Stuart. 1977. *The Dragon Flies of British Columbia.* B.C. Provincial Handbook No. 35. Victoria. 254 pp. Out of print.

Danks, H.V. (ed.). 1979. *Canada and its Insect Fauna. Memoirs of the Entomological Society of Canada* No. 108. 573 pp. (This useful work is a synthesis of the state of knowledge of the insect fauna of Canada, especially valuable for its extensive bibliographies of scientific works on insects, including annotated lists of various insect groups in British Columbia.)

Hatch, M. H. 1953-71. *The Beetles of the Pacific Northwest.* Parts I-V. University of Washington Press, Seattle. 2157 pp. (A detailed scientific treatise on the beetles of the region.)

Jones, J.R.J.L. 1951. *An annotated checklist of the Macrolepidoptera of British Columbia.* Entomological Society of British Columbia Occasional Paper No. 1. 148 pp. (The last published list of B.C. butterflies and moths. This publication is badly out of date and out of print.)

In addition, there are several excellent field guides available which deal with North America as a whole. These are in most libraries and bookstores.

*Robert A. Cannings*

## VEGETATION

A complex plant cover has developed in B.C. in response to the province's large size, geography (bounded by the Pacific Ocean and the Great Plains), varied topography, complex climatic patterns, and glacial history. Broadly speaking, B.C. is a cool, moist, mountainous, forested region. However, the province also has areas with Mediterranean-type, semi-arid, subarctic, and alpine climates. It has extensive plateaus, plains, and basins as well as several roughly parallel series of mountains. Forests dominate the vegetation but there are also extensive grasslands, wetlands, scrub, and tundra.

Several regional classification schemes have been used to order the diversity of B.C.'s natural environment. Divisions have been made based on climate, physiography, soils, and various combinations of these with plants and animals. This mix of factors has resulted in subdivisions such as forest regions, biotic regions, ecoregions, and biogeoclimatic zones. These schemes all have merit but for the purposes of this book they may be unnecessarily complex.

Therefore, in order to summarize the plant cover of the province, yet impart some appreciation of its vegetative diversity, it is convenient to subdivide B.C. into five major geographic regions and a sixth, primarily altitudinal region. They are Coast, Dry Southern Interior, Central Interior, Interior Wet Belt, North, and Alpine.

## COAST

The coastal region is predominantly mountainous, wet, and forested. Conifers dominate the plant cover, as they do over most of the province. The majority of the coastal forest is humid and is dominated at low to medium elevations by western hemlock and Western red-cedar. Douglas-fir and grand fir are abundant in the south; amabilis fir and Sitka spruce are more widespread in the north. Mountain hemlock, amabilis fir, and yellow-cedar or cypress predominate in the coastal subalpine forest. Deciduous tree species that invade and occupy disturbed areas include red alder, black cottonwood, big-leaf maple, and western flowering dogwood.

Mature coastal forest typically is dark and mossy, with numerous epiphytes. The understory is relatively sparse except in openings, and includes species such as salal, blueberries and huckleberries, devil's club, and various robust ferns.

The southwest Coast has a fairly extensive network of roads, and routes such as the Squamish Highway, Nanaimo to Tofino, and the North Island Highway to Port Hardy, provide good cross-sections of the regional vegetation. However, impressive old-growth forest with really big trees is almost all gone now except in a few preserved stands such as Cathedral Grove in MacMillan Park and the recently created provincial park on the Carmanah River.

The coastal region also includes the drier, rain shadow area of Georgia Strait and surroundings. This area enjoys a mild, Mediterranean-type climate and has unique vegetation. Arbutus (the only broad-leaved evergreen tree in Canada) joins Douglas-fir in much of the drier forest near the sea. Garry oak occurs with these two species but has a more restricted distribution. Garry oak also forms a distinctive type of deciduous woodland or savanna, in association with small pockets of vernal meadows and grassland, usually on warm, dry, rocky, south slopes. The meadows and grasslands are dominated by annual grasses and spangled with many showy, spring flowers such as camas, sea-blush, blue-eyed Mary, white fawn lily, and satinflower.

Wetlands are especially extensive along the outer coast north of Port Hardy. The wetlands (mostly bogs or coastal muskeg) form a mosaic with scrubby forests of cedar, cypress, western hemlock, and shore pine. The muskeg is widespread over the north coastal lowlands, which feature an unusual landscape of low hills covered with scrub forest and bonsai bogs. The main B.C. Ferry route north to Prince Rupert passes through much of this area as it threads its way through the archipelago-fjord terrain of the Inside Passage.

Maritime terrestrial vegetation occupies a variety of tidelands and nearby uplands between the forest and the sea. Rocky beaches are by far the most common type on the B.C. coast. Plant life is sparse and greatly influenced by tides and exposure to wind and salt spray. Some typical showy species of exposed rocky beaches are hairy cinquefoil, roseroot, mist-maidens, coast strawberry, and chocolate lily. Shingle beaches are fairly common and their upper reaches often support clumps of lime grass, beach pea, giant vetch, ocean strawberry, and springbank clover.

Sand beaches are uncommon and local, except for a few areas on Vancouver Island and the Queen Charlotte Islandsand Calvert-Hunter, Spider and Goose Islands. Vegetation is sporadic, but showy species of the driftwood zone include searocket, beach-carrot, and beach pea. Further up the beach may be silver bur-sage, big-headed sedge, dune grasses, sand verbena, paintbrush, and lupines. All of these beach types in exposed localities are backed up by a typical sea strip forest of Sitka spruce with dense salal underbrush.

*Shooting star*

Tidal marshes border the shorelines of estuaries and protected bays and inlets. Grasses and sedges dominate these marshes, but they also harbour showy species such as Pacific silverweed, springbank clover, and checker-mallow, as well as salt-tolerant oddities like glasswort and sea arrow-grass.

Probably the most spectacular areas to observe maritime vegetation are in Pacific Rim National and Cape Scott Provincial parks on Vancouver Island, and in Naikoon Provincial Park and South Moresby National Park on the Queen Charlottes. The Strait of Georgia also has numerous, more protected beaches and salt marshes.

## DRY SOUTHERN INTERIOR

This region covers south central B.C. from the Cariboo-Chilcotin district south to the International border, and includes the Okanagan Valley, the Thompson and Fraser plateaus, and drier parts of surrounding highlands and mountains, as well as the southern Rocky Mountain Trench below Golden.

Grasslands dominated by bunchgrasses (bluebunch wheatgrass and fescues, primarily) and semi-desert shrubs occupy valley bottoms and adjacent plateaus from the Riske Creek area south to the border. Similar but smaller grasslands occur in southeastern B.C. in the Kootenay and Columbia River valleys. Big sagebrush is the most common shrub, especially at the lowest elevations. Rabbit-bush and antelope-bush also dominate some scrub types. Showy species of this shrub-steppe include bitterroot, mariposa lily, prickly-pear cactus, sage buttercup, yellow bell, balsamroot, prairie smoke, scarlet skyrocket, wild flax, brown-eyed Susan, biscuit roots, milk-vetches, and fleabanes. The "pocket desert" of the Oliver-Osoyoos area has several species of restricted distribution, such as antelope-bush, prickly phlox, and pale evening-primrose.

Pristine grassland is difficult to find. Overgrazing is widespread and has resulted in invasion by aggressive herbs like cheatgrass and knapweed. Rapidly expanding cities and cropland have also destroyed much habitat. Major roads travel through disturbed habitat. Perhaps the best places to explore grasslands in relatively good condition are in the Douglas Lake country near Merritt, in the Lac du Bois area north of Kamloops, and on the Junction range southwest of Williams Lake.

Ponderosa pine and Douglas-fir dominate the dry forest, parkland, and savanna of the Southern Interior. Western larch is a common associate in southeastern B.C. These dry forests may occur above or below grasslands, but typically occupy well-drained benchlands and lower slopes. Douglas-fir, lodgepole pine, and trembling aspen dominate the forest of moister, intermediate elevations. Pinegrass forms the characteristic grassy understory of many of these stands.

Much of the plateau country (especially in the Cariboo-Chilcotin) is a parkland of open grassland and patches of aspen, lodgepole pine, and Douglas-fir. Numerous lakes and ponds dot the surface and are fringed by a variety of wetlands.

Higher elevation forests in this region are mostly continuous and coniferous, and are dominated by lodgepole pine, Engelmann spruce, subalpine fir, and, in the southeast, western larch. These forests are denser than those of the lower elevations, and typically have more shrubs and mosses and less grass in the ground cover.

## CENTRAL INTERIOR

The Central Interior of B.C. is a rolling upland with rounded hills and low mountains, largely consisting of the Nechako Plateau and the Fraser Basin. This broad, intermontane region is overwhelmingly forested. A mixed forest of lodgepole pine, trembling aspen, spruce, subalpine fir and less commonly Douglas-fir, black spruce, and paper birch covers the lower elevations in the major valleys. With increasing elevation, subalpine forests are dominated by subalpine fir, spruce, and lodgepole pine. Understory vegetation typically is well developed, with abundant shrubs and herbs as well as mosses. The relatively high, cold plateau country of the western Chilcotin and upper Blackwater districts has an extensive cover of lodgepole pine with minor white spruce and a sparse understory. Numerous marshes, fens, and wet meadows are a conspicuous feature of this area.

Alluvial forests of black cottonwood, often with a minor component of spruce, occur along the major streams and rivers. Wetlands are common in poorly drained, postglacial depressions or river oxbows. Small lakes and ponds are everywhere, while the northwestern part of the region in the lakes district is characterized by large lakes such as Babine, Ootsa, and Stuart. Natural grassland and shrub-steppe are common, occurring on some warm, dry, south slopes and ridges scattered in the valleys. Wetlands and dry ridges are often the best places in this region to see unusual vegetation and search for uncommon species.

Major parks that include a good sample of the regional vegetation are Carp Lake, northern Bowron Lakes, and northern Tweedsmuir.

## INTERIOR WET BELT

The Interior Wet Belt or Columbia-Kootenay region encompasses primarily the Columbia Mountains and the windward, southern Rocky Mountains from east of Prince George south to the U.S. border, as well as parts of the southern Rocky Mountain Trench. This is the wettest region in the interior of the province, and much of the vegetation, especially the forests, has a coastal character. Lower and middle elevations are primarily forested. Western hemlock and Western red-cedar dominate climax stands, but grand fir (in the extreme southern parts of the region), white spruce, Engelmann spruce, their hybrids, and subalpine fir are also common. Forests with a history of disturbance typically have western larch, Douglas-fir, western white pine, or black cottonwood. Lodgepole pine, trembling aspen, and paper birch also are common seral species except in the wettest parts of the region — where wildfires have been infrequent.

This region and the southwest Coast have the highest diversity of tree species in the province. Typical understory species include Douglas maple, yew, devil's club, false azalea, false-box, thimbleberry, blueberries and huckleberries, bunchberry, queen's cup, prince's pine, false sarsaparilla, and ferns. The wettest parts of this region, such as around Mica Creek and the upper McGregor River, have a dense undergrowth dominated by devil's club and ferns. Some mountainsides seem to consist of one continuous slope of devil's club.

Subalpine forests typically are dominated by Engelmann spruce and subalpine fir. Lopdgepole pine is abundant in some places, and even mountain hemlock is frequent in the wettest areas and those with the most snowfall, especially along the big bend of the Columbia River.

Several parks are located in the region, including Glacier, Mt Revelstoke, Mt Robson, Wells Gray, Valhalla, Kokanee, and Purcell Wilderness.

## NORTHERN BOREAL FOREST

North of roughly 56 degrees latitude is the land of the boreal forest in B.C. Forests predominate on the better-drained plateau, foothill,and cordilleran parts of the region. Lower and middle elevation forests are dominated by white spruce, black spruce, trembling aspen, and lodgepole pine, with lesser amounts of paper birch, subalpine fir, and balsam poplar. The prevailing poor tree growth reflects the short growing season and cold soil temperatures, which often are accompanied by poor drainage.

Poorly drained areas like the Fort Nelson Lowland support a mosaic of forests and wetlands. Black spruce bogs are the most common wetland type. Tamarack also grows in some bogs, but forms pure stands only in nutritionally rich fens and swamps. Tamarack swamps often have unusual species, including several orchids. Productive forests of white spruce and balsam poplar are common on floodplains, especially along the big rivers. Dry pine forests with a lichen ground cover are common on coarse outwash, whereas dense black spruce-feathermoss forests develop on imperfectly drained sites.

Within the boreal region, the Peace River district has a distinctive plant cover dominated by parkland or closed forests of aspen, white spruce, and some birch. Grasslands are locally common along the Peace River "breaks." Similar dry grassland and scrub occur, usually in small pockets, on steep, south-facing slopes above many of the major northern rivers -- most notably the middle Stikine around Telegraph Creek.

Subalpine environments in the North support open forests of white spruce and subalpine fir, grading with increasing elevation into parkland and eventually an extensive chest-high scrub of willows and scrub birch. The willow-birch scrub is especially widespread over the Stikine and Yukon plateaus. In some of the high, wide, subalpine valleys, a mosaic of shrubfields, wetlands, and Altai fescue grassland occupies the treeless valley floors, while a skirt of forest clothes the slopes above.

Representative parks include Spatsizi, Kwadacha, Atlin, Muncho Lake, and Stone Mountain.

## ALPINE

British Columbia is blessed with a widespread alpine zone that occurs on high mountains throughout the province. In southeastern B.C., alpine elevations start at about 2250 m; in the southwest, at 1550 m; in the northeast, at about 1400 m; and in the northwest, at about 1000 m.

Timberline trees are most often subalpine fir, Engelmann spruce, and mountain hemlock but also include white-bark pine, cypress, lodgepole pine, white spruce, and alpine larch. The alpine zone is, by definition, treeless, but trees are present although in stunted or krummholz form. The krummholz belt is usually fairly narrow, however, and alpine vegetation is made up primarily of shrubs, herbs, bryophytes, and lichens.

Low deciduous shrubs often dominate in the lower alpine, especially in the northern part of the province, where common species are willows and scrub birch. A dwarf scrub of prostrate woody plants is the most abundant form of vegetation in B.C.'s alplands. It predominates at the middle alpine elevations and is especially abundant in moister, snowier areas. In drier regions it tends to be confined to areas of snow accumulation, although some dwarf shrubs such as mountain avens are restricted to windswept, largely snow-free, ridgecrests. Important dwarf shrubs include the mountain-heathers, bearberry, crowberry, mountain avens, willow, and mountain blueberries and cranberries.

Alpine grass vegetation is also widespread. It becomes dominant at higher elevations of the drier regions, as in the south central interior, the Chilcotin district, and along the leeward slopes of the Rockies. In moister regions, grassy vegetation tends to be localized and often is restricted to steep south slopes or windswept ridges.

Meadows dominated by showy-flowered, broad-leaved herbs are common in the alpine, as they are also in the subalpine parkland that grades into the alpine zone. The parkland is an attractive mosaic of tree clumps and open, exposed areas of heath, grassland, and meadow. These mountain meadows are lush and florally diverse — the botanical glories of the high country. Typical species found virtually throughout the province include arctic lupine, arrowleaf butterweed, Sitka valerian, false hellebore, mountain daisy, cow parsnip, arnicas, cinquefoils, and pussy-toes. Glacier lilies and spring beauties lay out an early carpet in the meadows of the southern third of B.C.

At highest elevations, few species of flowering plants can survive. Most are cushion-plants or mat formers, such as moss campion, purple saxifrage, sandworts, and drabas. However, a few mosses, liverworts and numerous lichens persist and even thrive at the upper limits of vegetation. Plants, mostly lichens, occur over shattered bedrock, in fellfield and boulderfield, or as vegetation strips on frost patterned ground. Some of the lichen tundra, especially on limestone, is surprisingly colourful, and rich in species.

Most of B.C.'s major parks contain outstanding alpine areas, but good road access to these alpine gardens is limited to Manning, Mt Revelstoke, Kokanee, and Stone Mountain parks.

## FURTHER INFORMATION

*Biogeoclimatic Zones of British Columbia.* 1988. Research Branch, B.C. Ministry of Forests, Victoria, B.C. This is a colour map with detailed notes, photos and diagrams of the province, an excellent guide for both residents of, and visitors to, B.C.

Cowan, I.McT., and C.J. Guiguet. 1965. *The Mammals of British Columbia.* British Columbia Provincial Museum Handbook No. 11. Victoria, B.C.

Demarchi, D.A., R.D. Marsh, A.P. Harcombe, and E.C. Lea. 1990. The environment. pp 55-144 *in* R.W. Campbell *et al. The Birds of British Columbia.* Royal British Columbia Museum and Canadian Wildlife Service, Victoria, B.C.

Holland, S.S. 1976. *Landforms of British Columbia: A Physiographic Outline.* Bulletin No. 48 (2nd ed.). British Columbia Dept. of Mines and Mineral Resources, Victoria, B.C.

Krajina, V.J. 1965. "*Biogeoclimatic zones and biogeocoenoses of British Columbia.*" Ecology of Western North America 1:1-17.

Krajina, V.J. 1979. "*Biogeoclimatic zones.*" Map 23, Page 49 in A.L. Farley. Atlas of British Columbia. University of British Columbia Press. Vancouver B.C.

Meidinger, D., and J. Pojar (compilers and editors). *Ecosystems of British Columbia*: Special Ref. Ser. No. 6, British Columbia Min. of For., Victoria, B.C.

Munro, J.A., and I.McT. Cowan. 1947. *A review of the bird fauna of British Columbia.* Special Publication No. 2. B.C. Provincial Museum, Victoria, B.C.

Rowe, J.S. 1972. *Forest regions of Canada.* Publication 1300. Canada Dept. of Envir., Canadian Forestry Service, Ottawa, Ont.

Valentine, K.W.G., P.N. Sprout, T.E. Baker, and L.M. Lavkulich. 1978. *The soil landscapes of British Columbia.* Ministry of Environment, Resources Analysis Branch. Victoria, B.C.

*Jim Pojar*

## GEOLOGY

A geological map of British Columbia is an intricate jigsaw puzzle of thousands of tiny patches of different kinds of rock. For this reason, it is more convenient to use a map of physiographic regions (see Fig. 2) to describe the rock types of the area rather than a geological map. Each physiographic region contains a particular combination of rock types, topography, and landforms, so they provide a good basis for an account of the physical features that can be noticed while travelling around the province. Most of the mountain ranges, plateaus, and rivers that are named on this map are also identified on most road maps.

The geological complexity of British Columbia is largely due to its position on the western edge of the North American continent. During the past several hundred million years, while the North American Plate has moved slowly westward, it has, on several occasions, collided with microcontinents that were being carried eastward on the adjacent oceanic plate (see Fig. 1). The microcontinents were made up of various types of rocks — for example, some consisted of volcanic rocks like the present-day Hawaiian Islands. These landmasses became attached to the western edge of the continent and then were compressed into the long, narrow belts that are now the physiographic regions. Magma (molten rock) that rose up from the subduction zone during collisions either cooled and crystallized within the earth's crust, forming large plutons, or it erupted at the surface, forming volcanoes and volcanic rocks. Pre-existing rocks adjacent to plutons were heated and compressed, resulting in their recrystallization and internal rearrangement, and forming metamorphic rocks. Sedimentary rocks originated where detritus that had been eroded from uplands accumulated on lowlands and in shallow seas.

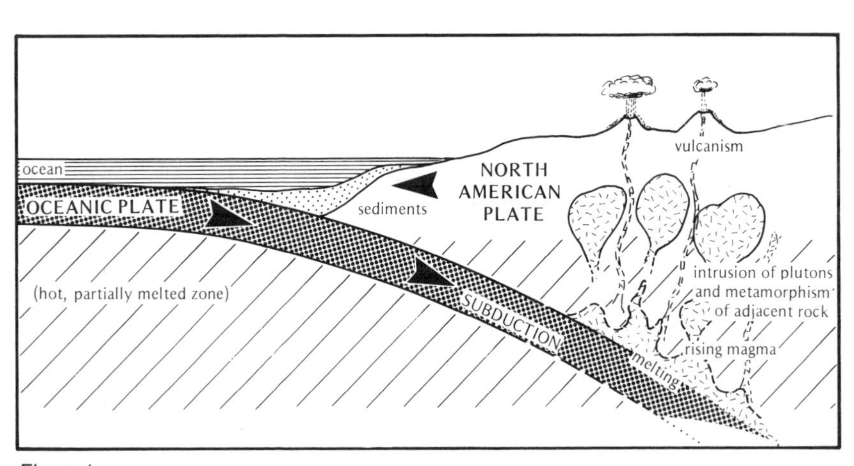


*Figure 1*

Over the course of millions of years, lowering of the land by erosion has counterbalanced the tendency of geological forces to raise the land. Thus what we see today at the

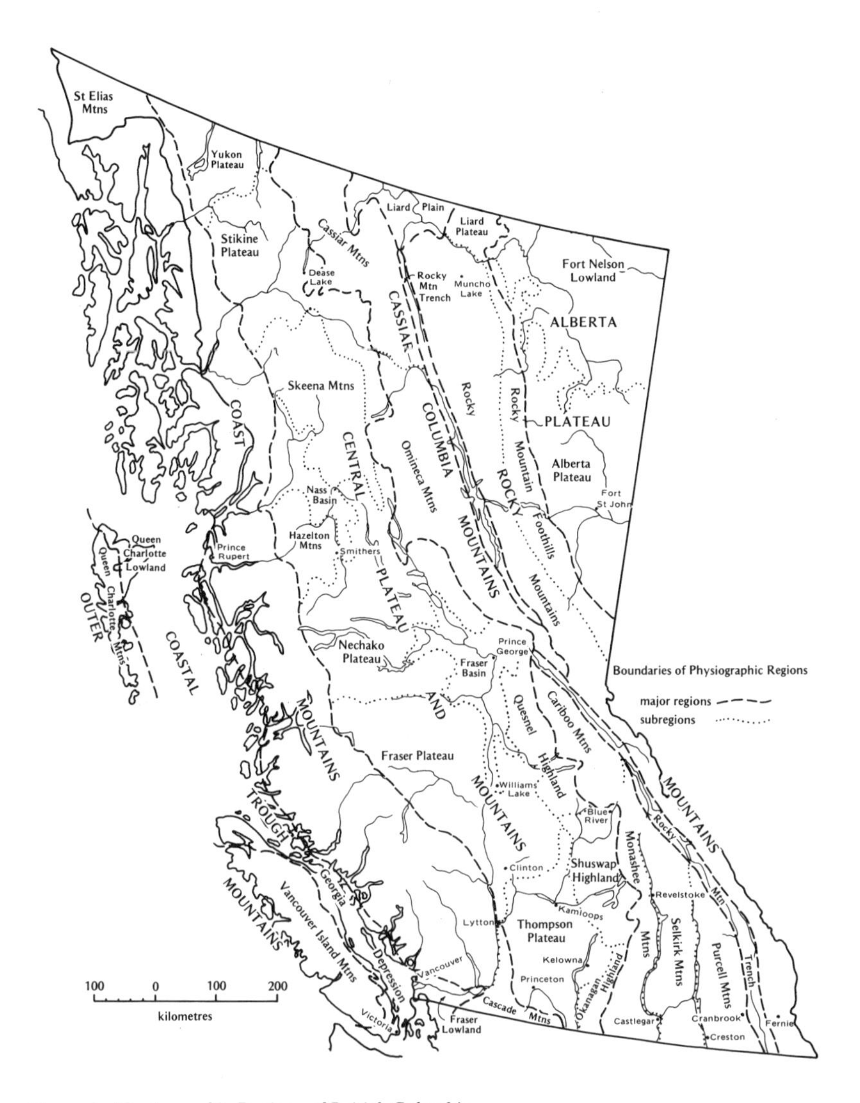


*Figure 2: Physiographic Regions of British Columbia*

earth's surface is an irregular horizontal slice through the rocks that were added, by one means or another, to the margin of the ancient continent.

The Outer Mountains and Central Plateau and Mountains physiographic regions each consist of two or more microcontinents. Their rock types include relatively old volcanic and sedimentary rocks, as well as younger plutons. Plutonic and metamorphic rocks are widespread along the margins of the old microcontinents within the present-day Coast and Cassiar-Columbia Mountains. The Rocky Mountains consist of old sediments that were thrust eastward onto the edge of the ancient North American continent by the force of the collisions. In the Alberta Plateau region, the ancient rocks of the continental core are buried beneath undeformed, relatively young sediments.

After our region had attained approximately its present configuration several tens of millions of years ago, there was a long period of quiescence, during which the land was eroded down to a low, rolling plain. Then volcanic eruptions occurred. Vast quantities of fluid lava poured out onto the plain, and low-lying areas were buried by lakes of lava. Subsequently, renewed tectonic activity resulted in uplift and warping of the erosion surface, with uplift being greatest along the present-day mountain belts. This initiated a new cycle of river downcutting and erosion. At the present time, the old erosion surface with its superimposed lava forms the dissected upland surface of the plateaus of the Central Plateau and Mountains region. Some remnants of the flat or gently sloping upland surface can also be identified in the Vancouver Island, Coast, and Cascade Mountains.

During the most recent episode of earth history, the landscape has been modified by the effects of glaciation. On several occasions during the past 2 million years, British Columbia has been inundated by the Cordilleran Ice Sheet, a huge complex of mountain ice fields and coalescing peidmont glaciers that extended from the Aleutian Islands in Alaska to the Columbia Plateau in Washington State. Nearly all the plateaus and lowlands in British Columbia were totally buried by glaciers, and even in the mountains, only the highest peaks protruded above the vast icy expanse.

Within the mountains, glaciation transformed the pre-existing valleys into cirques and U-shaped glacial troughs, and high ridges and peaks were sharpened into knife-edged arêtes and horns. Where the ice overrode lower ridges, plateaus, and lowlands, the land surface was smoothed and rounded by glacial abrasion. When the ice melted, most recently about 12,000 years ago, glacial sediments were deposited. Glacial till was laid down on all but the steepest slopes, and sands and gravels were deposited on plateaus and lowlands by meltwater streams. Some meltwater deposits were laid down on top of the melting ice or underneath it, resulting in the gravelly hills and ridges referred to as kames and eskers. Lakes were impounded by glacier dams during ice recession. Rock flour produced by glacial abrasion was carried into these lakes by meltwater streams and settled out on the lake floor, forming thick accumulations of silt and clay. Relative sea level was 100 to 200 m higher than now at the end of each glaciation, and so marine silts and clays, which commonly contain shells and stones carried by icebergs, are widespread in coastal lowlands.

The glacial features that were formed during the most recent glaciation are very apparent in the present-day landscapes. Glacial till forms a smooth blanket on most slopes where rocks are not exposed, and the material itself is visible in many road cuts. Meltwater sands and gravels and raised deltas are being excavated for aggregate in many gravel pits. Glacial lake sediments form extensive benches (the old lake floors) along many valleys and can be recognized by their pale (almost white) colour. Glaciomarine sediments blanket coastal lowlands. These latter two materials are highly susceptible to erosion and are commonly involved in landslides, creating problems of land management.

Since the ice disappeared, some parts of the landscape have been modified by river erosion and deposition. In some valleys, erosion has created terraces or deep canyons within the glacial valley fill. Elsewhere, depositional processes have resulted in the development of flood plains, alluvial fans, and deltas. Steep valley sides have been altered by the downslope movement of masses of rock and glacial materials, resulting in landslides, debris flows, and the formation of landforms such as talus slopes. Wind has eroded fine sand and silt from unvegetated areas such as flood plains and steep bluffs and redeposited the material as a thin layer of loess on gentle slopes.

Vulcanism has continued to play a role in landscape evolution until the present day. Volcanoes have been active periodically throughout the past 2 million years in three areas. The Stikine volcanic belt is a chain of volcanoes that runs north-south through Mount Edziza on the Stikine Plateau and includes the 250-year-old Aiyansh lava flow in the Nass River valley. The most recent eruption occurred in the Stikine belt and took place only about 50 years ago. The Anahim volcanic belt extends eastward from the ancient volcanoes of the Rainbow Mountains on the Fraser Plateau to the recent lava flows of Wells Gray Park. The Garibaldi volcanic belt stretches northward from Garibaldi Park for about 150 km. In each area, some eruptions occurred during glaciations, and the resulting interactions of molten rock and ice produced distinctive features such as flat-topped volcanoes, known as tuyas, and Black Tusk.

The present-day landscape — landforms, sediments and rocks — contains the evidence of its history. The story is there, waiting to be read by the observer who is willing to pause for a moment to think about the effects of tectonic forces, volcanism, glaciation, erosion, and deposition.

### *FURTHER INFORMATION*

Several booklets, maps and books on the geology of B.C. are available:

*The Identification of Common Rocks* (Free booklet available from: Public Distribution, Ministry of Energy, Mines and Petroleum Resources, Parliament Buildings, Victoria, V8V 1X4.)

*British Columbia Geological Highway Map* (Maps at scale of 1 cm to 12 km, geological information and long reading lists. Available from the above address.)

S.S. Holland. *Landforms of British Columbia, Physiographic Outline*. British Columbia Ministry of Energy, Mines and Petroleum Resources, Bulletin 48.

W.H. Mathews. *Garibaldi Geology*. Geological Association of Canada, Cordilleran Section.

G.H. Eisbacher. *Vancouver Geology*. Geological Association of Canada, Cordillaran Section.

*June Ryder*

# VANCOUVER ISLAND

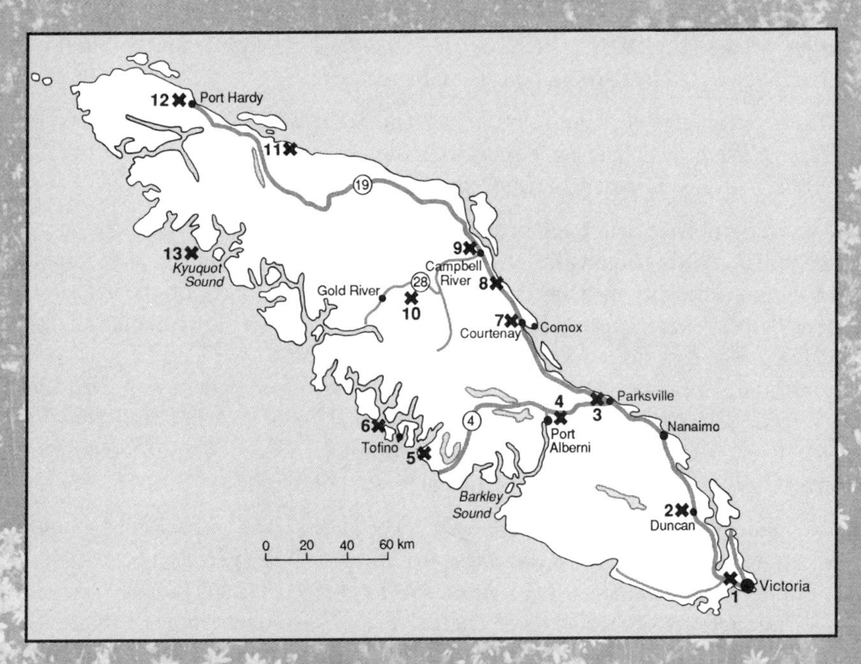


1. Victoria Region
2. Duncan - Cowichan Valley
3. Parksville and Qualicum Beach
4. MacMillan P.P. (Cathedral Grove), Port Alberni
5. Pacific Rim National Park
6. Clayoquot Sound and Maquinna P.P.
7. Courtenay - Comox and Forbidden Plateau
8. Miracle Beach P.P. and Mitlenach Island Park
9. Campbell River and Quadra Island
10. Gold River and West Side Strathcona P.P.
11. Johnstone Strait (North End) and Adjacent Islands
12. Port Hardy
13. Checleset Bay/Kyuquot Sound

# *VICTORIA REGION*

Set at the southern tip of Vancouver Island, Victoria, with its temperate climate and low rainfall, is a haven for naturalists. Over 320 species of birds have been recorded with about 250 reported every year. Offshore waters are rich with wildlife, including California and northern sea lions, and tidal sites have abundant crabs, sea stars, molluscs and many other invertebrates. Three pods of killer whales reside in the Puget Sound, Juan de Fuca Strait and the Strait of Georgia, and other transient pods pass through.

The forests contain large numbers of black-tailed deer which draw cougars into the area every few years. The land vegetation ranges from very dry upland to semi-rain forests, with a wide variety of native plants and imported exotics.

The mild, moist winters, relatively dry summers and diversity of topography and soils provide habitats for a wide selection of plants. Rainfall, year-round, varies considerably with an average of 76 cm, thus parts of the area are considered relatively dry. In summer, Victoria has the lowest rainfall for Canada. The east side of Vancouver Island including Victoria proper, Saanich Peninsula, Colwood, Metchosin and East Sooke, receives less rain than the west side. The Coastal Douglas-fir forest is found on the east side, and the west side is classified as part of the Coastal Hemlock forest. These two forests interweave in the Sooke-Jordan River area west of Victoria.

The geological history of southern Vancouver Island is complex. Rocks around Victoria and the Saanich peninsula range from Wark and Colquitz gneisses (heavily metamorphosed sedimentary and volcanic rocks of Paleozoic and possibly Triassic age) through volcanics and sediments of Triassic and Jurassic age. These are overlain by Cretaceous sandstones and shales and by Tertiary volcanics and sediments. Consequently the area has ranged from quiet deposition to intrusions and vulcanism. In addition, major faults cut through the area. The area consists of at least three separate "terranes" of over 200 large blocks transported from the south through major shifts of the earth's crustal plates. Blocks of Paleozoic and Mesozoic rocks were attached prior to or during Late Cretaceous time.

A visitor to the area should obtain the book *The Naturalist's Guide to the Victoria Region*, which provides a thorough look at the area and includes a bird checklist, descriptions of special birds, marine animals, mammals, plants, butterflies and an excellent picture of the complex geology of the region. Material for this section, except for the geology, was extracted from this book, and we thank the various contributing authors for their permission to use their information. Geological information was taken from the following:

"Lithoprobe-southern Vancouver Island: Cenozoic subduction complex imaged by deep seismic reflections," in the *Canadian Journal of Earth Science*, Vol. 24, 1987; the Geological Survey of Canada sheet, Map 1553; and *Victoria*, by J.E. Muller, 1980. Geology information for specific sites was provided by H. Paul Wilton, District Geologist, Southwest B.C.

With its wide diversity of habitats and warm, dry weather in summer, Victoria is famous for its birds. Early May is the peak time to see land birds with sometimes up to 120 species seen in a single, well-planned day. Fall is best for waterbirds. In 1987 and 1988 the Canadian record of 145 species for the Christmas Bird Count was set here.

**Birds:** (spring) violet-green swallows, brant, rufous hummingbirds, blue grouse, and yellow-rumped warblers; (summer) pelagic cormorants, glaucous-winged gulls, Heermann's gull, pigeon guillemots, red-necked phalaropes, marbled murrelet; (fall) Brandt's cormorants, turkey vultures, Vaux's swifts; (winter) waterfowl, glaucous gull, occasional gyrfalcons and peregrine falcons.
**Mammals:** eight species of bats, including little brown, yuma, and big brown; three species of shrew, deer mouse, Townsend's vole, red squirrel, mink, river otter, ermine, marten, black-tailed deer, wolves and black bears; Introduced Species: European rabbit, cottontail, gray squirrel, muskrat, black rat, Norway rat, house mouse and fallow deer.
**Marine mammals:** harbour seals, northern sea lions, California sea lions, killer whales, harbour and Dall's porpoises, and gray whales.
**Reptiles and amphibians**: Pacific tree frog, at least three salamander species, three kinds of garter snakes.
**Trees:** arbutus, Garry oak (the only native oak), Douglas-fir, broadleaf maple.
**Native shrubs:** Indian plum, red-flowering currant.
**Flowers:** satin flower, blue-eyed Mary, white lilies, chocolate lily and great camas.
**Butterflies and moths:** Satyr anglewing, mourning cloak, Mibert's tortoiseshell, cabbage white, Sara's orange tip, spring azure, anise swallowtail, northwest ringlets, black western tiger swallowtail, painted lady, purplish copper, woodland skipper,and red admiral, western lackey, snowberry bee hawk moth and Cerispy's eyed hawk moth and polyphemous moth.

The following sites of special interest within the greater Victoria area have been arranged in sequence. They begin on the west side of downtown at Ogden Point and head eastward around the peninsula (generally following the coast), inland and north to the ferry terminal, west along the inshore coast of the peninsula, south inland to Goldstream Park, and then south to Witty's Lagoon and East Sooke Park.

## Ogden Point Breakwater

At low tide, invertebrate animals typical of the exposed shore can be seen on the outside of this 0.5-km-long concrete pier, and those typical of more sheltered water on the inside. Pacific octopus may be seen occasionally between the granite blocks at low tide, and this is the approximate eastern limit of California mussels and goose barnacles.

**Marine life:** (outside) bull kelp, chitons, acorn barnacles, whelks, purple sea stars, California mussels, giant red urchins and goose barnacles; (inside) sea anemones, sea cucumbers, crabs, brittle stars, snails, sea urchins, and mottled and leather stars.
**Other animals:** sea birds, sea lions and harbour seals.

## Beacon Hill Park

Check the gulls carefully here as rare ones regularly appear. Goodacre Lake and other ponds on the northwest side of the park have numerous waterfowl in winter, and warblers pass through here in large numbers in fall and spring, particularly in the southeast corner of the park, at the junction of Cook St and Dallas Rd. The park has been developed as a horticultural garden, but look for the less cultivated sites to find remnants of native vegetation. The numerous rock outcroppings have been polished, grooved and striated by the last advance of the Pleistocene glaciers.

**Birds:** hooded merganser, wood duck, Eurasian wigeon, chestnut-backed chickadee, California quail, warblers, white-crowned and golden-crowned sparrows, surfbirds and black turnstone.
**Mammals:** raccoon.
**Marine life:** orange and white sea cucumbers, clams and jellyfish.
**Plants:** Garry oaks, Douglas-fir
**Geology:** amphibolite and amphibole Wark gneiss outcroppings.

The park sits along Dallas Rd south of, and adjacent to, downtown.

## Clover Point

This is one of the best places in the city for birds. A sewage outlet from the city lies off this point, and, as a result, numerous seabirds and gulls, such as Heermann's, gather to feed in the water just off shore. Tufted puffin is sometimes seen here. The point sits at the south (sea) end of Moss Street, east of Beacon Hill.

## Harling Point

Harlequin duck, black oystercatcher and black turnstone are found in and near the water near the middle of the point off the old Chinese cemetery. The rhinoceros auklet is sometimes also spotted in the water from here. At low tide the pools have good animal life. Also at low water, a north-dipping thrust fault, broken by numerous cross faults, can be observed. Metasedimentary schists of the Leach River complex have under-thrust schists of the Wark complex. This fault represents one of the most important sutures between two of the "terranes" described earlier. The point lies half way between Beacon Hill Park on the west and Victoria Golf Course on the east.

## Sea end of Bowker Avenue

Every Tuesday morning throughout the year, at 9:00 AM, a sub group of the Victoria Natural History Society meets here to check things out and then go to some other site in the region for a half to full day. Visitors are welcome. Bowker is reached at the east end of Fort, just after it becomes Cadboro Rd.

## Cattle Point and Uplands Park

The outcrops at Cattle Point are typical of the Colquitz gneisses. They are more strongly banded and lighter in overall colour than the Wark rocks. These Colquitz deposits are probably derived from Paleozoic sedimentary rocks metamorphosed in the Jurassic.

**Birds**: waterfowl, shorebirds, gulls, harlequin duck; (Garry oak forest) California quail, Bewick's wren and bushtit; (upper areas) killdeer and savannah sparrow.
**Plants:** Garry oak forest, straight beak buttercup, yellow violet, upright chickweed and tall woolly-heads.
**Marine life:** shrimp, chiton, crabs and the six-armed starfish.

The park and point lie east of the downtown, along Beach Drive.

## Ten Mile Point

Cadboro Bay and the point are excellent spots to find rocky shore creatures. The islets at the very end of the point are part of an Ecological Reserve so there is no collecting of sea life on or near them unless you have a permit. Sea mammals are often seen in Cadboro Bay just south of the point.

Ten Mile Point is underlain by intrusive granitic rocks believed to be of Jurassic age.

**Birds:** marbled and ancient murrelets, migrant upland birds.
**Marine life:** barnacles, several species of limpets and periwinkle snails, anemones, porcelain crab, hermit crab, red rock crab, orange and white sea cucumbers and chitons of various kinds; (seaweeds) bladderwrack, sea cabbage and leafy brown kelp.

The bay and point lie on the east side of the region, east of the University.

## University of Victoria

The native woodlands and exotic gardens found on campus provide some good birding opportunities.

**Birds:** (residents) pileated woodpeckers, Bewick's wren and bushtit; black-headed grosbeak, Hutton's vireo, western screech owl, Cooper's hawk and band-tailed pigeon.

## Swan Lake - Christmas Hill Nature Sanctuary

The lake and adjacent shores host a wide selection of freshwater birds, while nearby grasslands are home to upland birdlife. Moist spots are home to Vancouver Island beggarticks, a plant known only from the island. The summit of Christmas Hill is part of the Swan Lake Sanctuary. The trail up begins at the head of Woodhall Crescent, west off Quadra Street, north of Swan Lake.

**Birds:** ruddy duck, marsh wren.
**Mammals:** muskrat, river otter.
**Plants:** Douglas-fir forest, poison hemlock; (Christmas Hill) Garry oak.

Swan Lake Sanctuary lies three blocks south of McKenzie, just east of Douglas Street (Hwy 17). Take Ralph east off Douglas and then south on Swan Lake to the headquarters and nature centre. Information on the sanctuary, trail maps and check lists, are available at the centre. In addition they have displays and conduct a wide variety of programs indoors and out with organized birding walks held most weekends.

## Blenkinsop Lake and Observation Tower

The Victoria Natural History Society has erected a tower to overview the lake and vegetation on this site.

**Birds:** green-backed heron, wood duck, black-headed grosbeak and yellow warbler.
**Mammals:** river otter and mink.

To reach the site, find Blenkinsop which runs north-south between McKenzie and Royal Oak. Take Blenkinsop to Lohbrunner Rd, and go west to Lochside Trail which leads to the observation tower.

## Mount Douglas Park

Eared grebes occur offshore in winter, one of the few places in the region to see them. Along the road to the summit look at plants of the Douglas-fir forest; in the summit area grassy-bald vegetation is found.

The bedrock in Mount Douglas Park consists of Colquitz gneiss. Watch for small openings in the rocks where early pioneers appeared to have "mined" rusty shear zones, presumably looking for gold. Mt Douglas, Mt Tolmie, and other prominent hills around Victoria are monadnocks, meaning that they often stood as "islands" above the Pleistocene ice sheets that covered most of the land.

**Birds:** pileated woodpecker, eared grebes, northern rough-winged swallow, Steller's jay, varied thrush and red crossbill.
**Mammals:** red squirrels and raccoons.

The park lies almost due north of downtown Victoria. Shelbourne runs north and intersects Cordova Bay Rd in the park, providing direct access. The road to the top, with an excellent view of the city, takes off from Shelbourne, just after it enters the park from the south.

## Quick's Bottom

The Victoria Natural History Society has erected an observation tower at the edge of this shallow pond.

**Birds:** gadwall, blue-winged and green-winged teals, marsh wrens, red-tailed hawk and the occasional Lincoln's sparrow.

To reach the spot take Royal Oak Drive west to West Saanich Rd and north on West Saanich to Markham. Turn left on Markham for a short stretch to a very small roadside parking site next to the entrance to a path. Take the trail to the observation tower.

## Elk/Beaver Lake Park

The two lakes, now joined, contain a good selection of birds and aquatic plants, with the lakeshore having wet woodland species.

**Birds:** hooded and common mergansers, American coots and sometimes trumpeter swans.
**Mammals:** river otter, mink and red squirrels.
**Amphibians**: red-legged frog, western painted turtle.
**Trees**: Douglas-fir and cascara.

The lakes sit on the west side of Hwy 17, north of the intersection with Royal Oak Drive. A hiking trail circles both lakes.

## Bear Hill Park

Rising one km northeast of Elk Lake, this park contains dense woods with varied thrushes and rufous-sided towhees.

Bear Hill is a knob of resistant granodiorite of Jurassic age. On the southwest side of the hill, the granitic rock can be seen to intrude volcanic rocks.

**Habitat:** (on top) Garry oak forests, bald grassy areas; Douglas-fir forest.
**Flowers:** satin flowers, blue camas, sea-blush, canary violet, spotted coral-root, striped coral-root, tiger lily and ocean spray shrub.

Take Keating Cross Rd west off Hwy 17 to Oldfield Rd and turn left (south) to Bear Hill Rd. Turn left on Bear Hill and follow it east, then south to the park.

## Martindale Flats

The flats are one of the best places in the region to find a gyrfalcon and/or peregrine falcon hunting the numerous waterfowl that winter on these wet flooded flatlands. Scan the few large trees for birds of prey, and the wet areas for waterfowl. Over 90 species of birds are found here on Christmas bird counts.

**Birds**: waterfowl including trumpeter swan and occasional tundra swan, gyrfalcon, peregrine falcon, white-crowned sparrow, golden-crowned sparrow and Eurasian skylark. (Note: Skylarks are heard and seen at Victoria airport from the end of February to the end of July; also north of Mount Newton Crossroad to Central Saanich Rd near a large greenhouse/warehouse complex.)

To reach the flats, turn east off Hwy 17 (Patricia Bay Hwy) onto Martindale Rd and drive to the nearby agricultural flats. Alternatively turn off Hwy 17 onto Island View Rd.

## Island View Beach and Cordova Spit

Waterfowl in large numbers occur along the shore in winter. The fields on both sides of Island View Rd are places for Eurasian skylarks from March to July.

To obtain a good overview of the vegetation and life of the area, walk from the parking lot at the beach north to the spit.

**Birds:** three loon species, red-necked grebe, common murre, brant, great blue heron, shorebirds, Heermann's gull (late summer, fall), all three scoter species and old squaw.
Sea mammals are often seen off the spit.
**Plants:** gold star, western crabapple, Nootka rose, beach morning glory and jaumea.
**Marine life**: purple shore crabs, gunnel or prickleback fish, acorn barnacle and blue mussel.

Take Island View Rd east off Hwy 17, just north of the East Saanich turnoff. An alternate route lies to the north along Mount Newton Cross Rd which leads into Church Rd and on further towards the spit, through the Indian Reserve.

## John Dean Park

This park straddles the summit of Mt Newton and, like Bear Hill, consists of a resistant knob of granodiorite of Jurassic age, which was polished and rounded much later by the advancing Pleistocene glaciers.

**Plants:** vanilla-leaf, white trillium, red-flowering currant, Douglas-fir forest

From McTavish Rd south of the airport, take East Saanich south through the Experimental farm to Dean Park Rd. Take Dean Park west to the park. Leave the car in the lot near the top and explore the site.

## Tsehum Harbour

This harbour, just south of the ferry terminal, has extensive mud flats at low tide making it a good place for shore birds and, in winter, ducks, loons and grebes. Nearby woods contain varied thrushes and band-tailed pigeons. To access the harbour take Mcdonald Park Rd right off the highway and then right onto streets such as Resthaven or Blueheron to reach near the water.

## Deep Cove

This secluded bay is sometimes one of the first spots where vagrant offshore bird species appear. Scan the resident waterfowl for specialties. A rich invertebrate life is exposed at low tide.

**Marine mammals**: sea lions, harbour seals.
**Marine life:** horse clam, Japanese clam, native little neck clam, bent nose clam, 15 species of sand worms including Pacific lugworm, dungeness crab, horse crab, kelp crab, coonstripe shrimp and sand shrimp.

West Saanich Rd skirts along the shore providing numerous opportunities to pull over, stop and look.

## Thetis Lake Park Nature Sanctuary

In spring this is one of the best spots to see the first yellow-rumped warblers hawking insects along the lake shore, and it is the only known site on Vancouver Island for broad-leaf arrowhead. All of the rock outcrops in Thetis Lake Park consist of Wark gneiss, probably derived from the metamorphism of Triassic volcanics.

**Mammals:** red squirrel, raccoon, mink and river otter.
**Habitats:** Douglas-fir, Garry oak forests, and grassy-balds.
**Plants:** golden-back fern, wild Clarkia or farewell-to-spring, white meconella (a native poppy), the parasitic one-flowered cancer-root, purple snake-root, and Sierra snake-root.

The park is accessed off Hwy 1 a short distance west of the junction with 1A.

## Freeman King Park, Thomas Francis Park, Nature House

The trails here are excellent for birds in May and June. One trail is good for wheelchairs.

**Habitats:** Douglas-fir forest, wet forests, and Garry oak forest.
**Plants:** California tea and slender woolly-head.
**Birds:** turkey vultures, Cooper's and red-tailed hawks, blue grouse and bald eagle.

Prospect Lake Rd runs along the east side of the park and Munns Rd takes off from Prospect Lake Rd north west through the park.

## Goldstream Provincial Park and Nature Centre

The park has a wide selection of hiking trails to explore and an excellent collection of native vegetation. Across the bridge at the picnic area look for Pacific waterleaf.

The rocks outcropping throughout Goldstream Park are schists and slates belonging to the Leech River Complex, one of the exotic "terranes" which became attached to Vancouver Island by underthrusting along the Survey Mountain Fault during the early Tertiary period. Rocks exposed on the upper part of Mt Finlayson belong to the Wark gneissic complex.

**Birds:** American dipper, varied thrush, red-tailed hawk, rufous hummingbird; (Mount Finlayson) turkey vulture, bald eagle, western bluebird, violet green swallow and Steller's jay.
**Mammals:** red squirrels and raccoons.
**Fish**: Excellent.salmon spawning site in fall!
**Plants**: crinkle-awned fescue, old Douglas-fir, Western red-cedar, black cottonwood, broadleaf maple, grand fir, wild ginger, Fritillaria lilies (some may be hybrids of black lily and chocolate lily) and gold-star.

The park is accessed along Hwy 1, west from Victoria just before the road begins the climb to the Malahat.

## Fort Rodd Hill National Historic Park

Masses of waterfowl and other water birds gather in Esquimalt Lagoon adjacent to the park in winter. One of two known recent nesting sites in Canada for the western purple martin is found on a ship in drydock at the nearby naval base. Watch for these birds hawking insects over the lagoon from May to August. This site is excellent for viewing black-tailed deer.

**Birds**: brant, herons, gulls, shorebirds, purple martin and migrating warblers.
**Mammals:** red squirrel, cottontail rabbit, and black-tailed deer; (Esquimalt Lagoon) river otter and mink.

The park is accessed off Hwy 1A a short distance south of the intersection with Hwy 1.

## Witty's Lagoon Park, Nature House and Tower Point

Water and shorebirds use this quiet lagoon most of the year.

The beach is mostly sand and is occupied by numerous clams and other burrowing creatures. Moon snails make "sand collars," nearly complete circles of sand and mucus, to hold their eggs. If found, they should be left undisturbed. You can also look for shrimp burrows, which have mounds of sand around the opening.

Pillowed basalt lava flows are exposed in the lower reaches of Metchosin Creek. These structures commonly develop when lava is extruded directly into water. These basalt flows belong to the Metchosin terrane and were erupted during the Eocene epoch. To find these unique structures, said to be one of the best exposures in North America, park the car beside Olympic View Drive near the field and walk to the west side of Tower Point and the first pocket beach.

**Birds:** ospreys, bald eagles, grebes, cormorants and shorebirds.

At the exit of Hwy 1A, take it (Hwy 1A) to Jacklin Rd and then to Metchosin Rd to the Park. The main beach is reached by parking at the end of Witty's Beach Rd.

## Matheson Lake Park

Woodland birds are found here all year, and a lakeside trail passes through Douglas-fir and arbutus wood.

**Birds:** golden crowned kinglet, chestnut-backed chickadee, pileated woodpecker, golden eagle, and Steller's jay.
**Plants:** yellow skunk cabbage, Western red-cedar, and gnome plant (a saprophyte).

The park lies just off Rocky Point Rd east of the Sooke Basin.

## East Sooke Park

Hundreds of turkey vultures and hawks collect here in late September before taking off across Juan de Fuca Strait on migration.

East Sooke Park is underlain by various types of gabbro of Eocene age, known collectively as the Sooke Gabbros. These rocks together with the overlying Metchosin basalt flows, constitute a slab of oceanic crust of Eocene age which was underthrust and fused against the southern end of Vancouver Island in the mid-Tertiary.

**Birds:** barred owl, pygmy owl and great horned owl, Hutton's vireo, loons, grebes, cormorants and mergansers.
**Plants**: yellow skunk cabbage and arbutus.
**Marine life:** California mussel, dog whelk, porcelain crab, tar spot sea cucumber, sea worms, purple sea urchin and giant green anemones.

The park lies near the southern tip of the Island, south of the Sooke Basin.

## Whiffin Spit

The long thin spit nearly stretches across the lagoon. The wide expanse of intertidal zone is exposed at low tide and is rich with marine invertebrates. This is the approximate easternmost penetration of many exposed-coast species of invertebrates.

**Birds:** shorebirds, black scoter and American black oystercatcher.
**Marine life:** sponges, compound sea-squirts, bryozoa (moss animals), sea slugs, sea lemons, leopard nudibranch, porcelain crabs, peanut worms, and sea anemones (*Urticina coriacea*, and *U. crassicornis*).

To reach this spot drive two km west of Sooke on Hwy 14 and turn left at the sign for Sooke Harbour House. There is a parking lot where the road reaches the spit.

## Sooke River Potholes Park

The Sooke River potholes were formed by the "drilling" action of boulders swirling around in the river current. Boulders become trapped in whirlpools by the current and literally drill holes into the bedrock, which here consists of pillowed Metchosin basalt. This is the only known location of Sierra wood-fern in B.C.

To reach the site drive north from Miles Landing, off Hwy 14, along the east side of the Sooke River.

## Botanical Beach

This is the first accessible site west of Victoria where typical exposed west coast plants and marine invertebrate animals may be seen in numbers. Excellent tide-pools occur along the sandstone beach, and forests are primarily western hemlock.

**Marine life:** Green anemones, purple urchins, limpets, chitons, barnacles, pink coralline algae and sea palms (*Postelsia*); occasionally small octopi.
**Marine mammals**: gray whales and killer whales.

To reach the site, take Hwy 14 from Victoria, via Sooke and Jordan River, to Port Renfrew. Just before reaching the Port Renfrew Hotel and Government Wharf, turn left on a gravel road (Cerantes Rd). Follow this road as far as possible. It has become very rough and overgrown in recent years. Follow the branch up to the left to reach the most interesting section of shore. The right branch leads to the western end of the beach, and one can do a circle tour by hiking along the beach and returning by the other road.

*Phil Lambert*

## The Ferries

The Tsawwassen (Vancouver) - Swartz Bay (Victoria) ferry route is a must ride for birders with a very wide diversity of water birds in March, April and May.

This route through the Gulf Islands between Tsawwassen and Swartz Bay, affords good views, particularly through Active Pass, of many shoreline exposures of Cretaceous sedimentary rocks. These conglomerates, sandstones, and shales are completely unmetamorphosed and only gently deformed. At the Swartz Bay terminal, there is an excellent exposure of Cretaceous shales and sandstones alongside the parking lot. The beds dip at about 45 degrees which is unusually steep for the Cretaceous sediments in the area.

Ample parking is usually available at both ends, making a return trip as a foot passenger possible.

**Birds:** bald eagle, northwestern crow, glaucous-winged gulls, common tern, red-necked phalarope, Bonaparte's gull, Heermann's gull, (possibly) parasitic jaeger, Pacific loon, Brandt's cormorant, common murre, ancient murrelet, marbled murrelet, common loon, surf and white-winged scoters, both goldeneyes, oldsquaw and mergansers.
**Marine mammals**: harbour seals, Steller's sea lions, killer whales and harbour porpoises.

The ferry return run from Victoria to Port Angeles is not as good for birds, but several species usually not seen on the Active Pass Route can occur while crossing the Juan de Fuca Strait. This is the best chance to see some of the more pelagic species such as sooty shearwater, northern fulmar, and possibly kittiwakes and Cassin's auklet. Like the other trip, summer is not very productive. The best times are March, April and May. Fall and winter are fair birding times. The best spot for waterfowl is near the harbour at Port Angeles.

**Birds:** marbled murrelets, rhinoceros auklets, sooty shearwaters, brants, Cassin's auklets, parasitic jaegers, western gulls and tufted puffins.
**Mammals:** Dall's porpoise.

*David Sterling*

**Further Information**
By far the best all around reference is *The Naturalist's Guide to the Victoria Region*, edited by Jim Weston and David Stirling, and sponsored by the Victoria Natural History Society, 1986. It is available from the Society at P.O. Box 5220, Station B, Victoria, B.C. V8R 6N4 and local bookstores. • Another book worth purchasing is put out by the Sierra Club of British Columbia, *Victoria in a Knapsack: A Guide to the Natural Areas of Southern Vancouver Island.* This publication adds human history and maps for each site, which alone are worth the price of $7.95. Other features include detailed ferry information, plus encapsulated looks at history, landforms, weather, sea life, plants and tips on using the lands around the island. It may be obtained from the Sierra Club of Western Canada, P.O. Box 202, Victoria, B.C. V8W 2M6. • Local contacts include the Victoria Visitor Bureau at 812 Wharf St., Victoria, V8W 1T3. Phone (604) 382-2127; B.C. Parks Malahat District, 2930 Trans Canada Hwy, Victoria, B.C., V9B 5T9, (604) 387-4363. Capital Regional Parks, 490 Atkins Ave., Victoria, V9B 2Z8. Phone (604) 478-3344; Ministry of Forest Recreation, 2nd Floor, 610 Johnson St., Victoria, B.C., V8W 3E7, phone (604) 387-1946. Royal British Columbia Museum, 675 Belleville St., Victoria, B.C., V8V 1X4, phone (604) 387-3014. All of the above agencies have information and should be contacted for available printed materials. The Royal B.C. Museum, just east of the parliament buildings, has excellent natural history exhibits and one of the best bookstores for information in the province. Other contacts: Victoria Natural History Events Tape - 479-2054; Rare Bird Alert - 592-3381. Philip Lambert, Royal British Columbia Museum; David Nagorsen, Royal British Columbia Museum; Leon E. Pavilic, Royal British Columbia Museum; David Stirling; Jeremy Tatum; H. Paul Wilton, B.C. Geological Survey.

# DUNCAN - COWICHAN VALLEY

The Duncan area, just over the Malahat from Victoria, offers marshes and tidal flats for birds all year and the opportunity to see the Vancouver Island marmot and the last herd of Roosevelt elk on the south part of the Island. Spring flowers are prolific in certain spots with the flowering period extending from February to June. The best spots in the area include the Cowichan River Foot Path (fishing and birds), Shoal Islands (birds and flowers), Mt Richards (birds and spring flowers), Somenos Marshes along the highway (winter and spring birds), Lake Cowichan (elk and fishing), and Heather Mtn (marmots). Mt Tzouhalem is excellent for western bluebirds and spring flowers from early April through June.

## Cowichan Estuary

Lying adjacent to and southeast of Duncan, the estuary is good for birding all year with particularly attractive spring and fall migrations.

**Birds:** waterfowl, shorebirds and falcons.

Access is from Tzuhalem Rd on a private industrial road alongside the CN right of way. Leave the car in the parking area on the right, a short way in to the estuary. Hike the road or alongside the right of way out to the private dock area. Watch for lumber trucks on week days.

### Cowichan River Footpath

This 19-km all-seasons trail runs along the river adjacent to the tracks and begins southwest of Duncan. The trail passes through a wide variety of habitat including old-growth rain forests of fir, hemlock and cedar with huge sword ferns. The area is good for birding and flowering plants (in the spring through June). It takes about six hours of good fast walking to do the route one way to Skutz Falls. If time is short, take the Holt Creek 2.5 km loop that cuts back at the Creek, about a kilometre west of the trail head.

To access the trail from Duncan drive west over the bridge on Allenby to Indian Rd, then southwest on Indian to the cross roads intersection of Glenora and Indian. Glenora changes to Vaux at that point. Follow Vaux, which changes to Robertson, until it terminates at the parking lot at the trail head. A good map of this access route and trail guide is published by the Cowichan Fish and Game Association, available from the Tourist Office in Duncan.

### Hill 60

This open, dry, rocky site lies west of Duncan and is good for spring plants and birds.

To reach the site drive 13 km west from the Island Hwy along the Lake Cowichan Hwy. At the end of the three-lane road make a sharp turn north (right) onto a gravel road and small parking area. From the parking site choose your route up the hill to the ridge where it is open along the bluff to the triangulation point.

### Mayo Lake

There is good spring birding along the trail that goes part way around the south side of the lake. Access the spot west along the Lake Cowichan Hwy. Watch for the Lake Mayo and Shutz Falls turnoff which is a paved road.

### Lake Cowichan Area

This region, around the lake, is great for birding and for flowering plants in the spring and early summer. The Shaw Creek valley, on the northwest side of the lake, just east of Heather Mtn, provides an opportunity to see the last elk herd left on the south Island. Take the main road west along the north side of the lake to the Pacific Logging Co. road and follow it north up the valley.

### Heather Mountain

Some of the best alpine flower meadows to be found on the south Island are here, together with an active colony of Vancouver Island marmots. To find the spot drive to the far northwest corner of Lake Cowichan.

Continue west on Line 1 for about 1.5 km. Turn sharply right, or north, and drive through the gate and 6.5 km up a long hill. Turn sharply right again just before a bridge and drive a short distance to a small parking spot. B.C. Forest Products have erected large green signs directing visitors to the Heather Mtn Trail. Roads are closed to traffic during working hours. A fairly well-marked trail of about 40 minutes walking time leads to a small lake at the top of a ridge, the beginning of an alpine meadow full of flowers. After another 20-minute walk you reach a saddle where the trail divides and becomes unclear. From here to the summit is a tough steep climb only for experienced hikers or those led by an experienced person.

### Mt Whymper

This mountain is also a good area for alpine flowers and contains a small Vancouver Island marmot colony. It lies on the north side of the Chemainus River and is difficult to find in the complex forestry road system of the area. Contact local naturalist to help find this site.

### Sutton Creek Wild-Flower Ecological Reserve

Well known for the spring (mid-April) flowers, this spot lies immediately west of the community of Honeymoon Bay, on the west side of McKenzie Bay and on the south side of Lake Cowichan. At Duncan take Hwy 18 west through Honeymoon Bay, turn left off the Gordon Bay campsite road, and follow the signs to the parking area.

### Cherry Point Beach

Lying south of Duncan and east of the community of Cobble Hill and the Island Hwy, this beach is very good for winter birding, particularly waterfowl out on the bay. To access the beach take Fisher Rd east from the Island Hwy at Cobble Hill, to Cherry Point Rd and follow Cherry to Garnet Rd. Take Garnet to the sandy beach.

### Cobble Hill

The walk up and along the top of this hill is very good for flowering plants and birding, and the site is part of the Vancouver Island Plantation Forest Reserve. For the easiest access route follow Thain Rd northwest from the community of Cobble Hill, continuing as it swings southwest. At the powerline crossing, look for the small parking area on the east side (left) just after Thain swings southeast. Park and walk up the fairly easy slope. The trail begins as an old jeep road and then becomes a foot trail to the top. The walk takes up about an hour.

### Mt Tzouhalem Ecological Reserve

A great place for flowers in March through June, and one of the few spots on Vancouver Island where western bluebirds nest (on the flat-topped hill). The site is considered the most outstanding example of Garry oak-spring wildflower habitats on Vancouver Island and is home to six endangered species. The site lies immediately east of Duncan. On the Island Hwy from Victoria, drive north over the Silver bridge (south side of Duncan) and turn east at the first traffic light. Head east a short distance and then angle northeast and east (generally east) onto Maple Bay Rd. Turn right off Maple Bay Rd on to Kingsview Rd, and then right on to Balcarra Rd and left on to Chippawa Rd. Follow Chippawa a short distance to its end and the small parking lot. Leave your car and take the trail into the reserve and up to the top of the flat-topped hill and viewpoint.

## Quamichan Lake

Lying immediately east of Duncan, this lake, with an island and surrounding marshes, is a great birding place. The site is reached by driving east from Duncan on the Trunk Rd which turns into Tzouhalen Rd. Turn north (left), shortly after you cross Quamichan Creek, on to Maple Rd and continue north on Indian Rd when Maple swings northeast. Follow Indian straight north to Art Mann Park at the lake. A boat ramp is available at the park.

## Somenos Lake

The marshes, small bodies of water and lake, are one of the best places on Vancouver Island to view water birds in the winter. Ducks occur in the hundreds. At the Somenos Centre an observation blind has been constructed to provide easy viewing, but good winter viewing can be had right from the highway.

**Birds:** Canada geese, cinnamon teal, marsh wren and black-headed grosbeak.

To reach the marsh go east off the highway at the traffic light just north of the Silver Bridge and take Maple Bay Rd a short distance. Turn left (north) at Lakes Rd and park at Somenos Creek bridge at the bottom of the hill. Hike west (left) along the creek to the marsh. The lake and outer marshes require a boat. A public boat ramp lies at the east end of Drinkwater Rd which runs east from the highway north of the Forest Museum entrance. Nearby Richard Creek marsh also provides good birding. Take Mays Rd east and park at the Herd Rd bridge. It is also possible to reach this marsh by boat up the creek from Somenos Lake when the water is high.

## Mt Richards

This area offers spring flowers and year-round birding opportunities. Richards Trail is good for birding along the top end of the valley. To reach the site, which is north of Duncan, leave the highway at Stratford's Crossing and take Westholme Rd for a kilometre. Then turn on to Richards Trail and drive southeast for another kilometre and watch for the trail on the east (left) side of the road where the rock bluffs come down to the road. Park and take the game trail up the slope, which is about a 300-metre climb. It takes about an hour of easy walking to the top.

## Shoal Islands and Chemainus Estuary

Located still further north and east of the highway, the marsh and tide flats provide good birding all year, with an especially good winter collection of waterfowl. The Islands have one of the best stands of Rocky Mountain juniper on Vancouver Island, and spring wildflowers are great from February to June. Access is east off the Island Hwy towards Crofton, but instead go north to the beach by way of the B.C.F.P. pulp mill gate. Check in at the gate and hike the abandoned railway fields and tide flats. The causeway is closed to walking during working hours. There is an open picnic site on the point.

### Bare Point, Chemainus Bay

Cormorants nest at this spot which is also good for winter birding. To reach the site take Power House Rd off Chemainus Rd (old highway) south of the community of Chemainus.

### Chemainus Lake

The lake provides good birding, especially during spring and fall migration. Hike the old road to the left before you reach the parking area. The lake lies immediately west of the highway. Turn left on to the logging road just north of the stop light at the Chemainus turn off. Follow the logging road about 0.4 km to a stop sign. Park on the right.

**Further Information**

Local maps are obtained from the Chamber of Commerce Tourist office in Duncan. Topographical maps include the 1:50,000 sheets Duncan 92 B/13 and Cowichan Lake 92 C/16 which are available at Island Blue Print, 905 Fort St., Victoria V8V 3K3, Phone 385-9786.A good contact to get in touch with is the Cowichan Valley Naturalists Society, Box 361, Duncan, B.C. V9 L 3X5. Phone the Ecomuseum for additional assistance at 746-1611.

*Syd Watts*

# PARKSVILLE AND QUALICUM BEACH

Habitats in the Parksville - Qualicum Beach range from alpine to seashore here in the central island. The highlights are the Vancouver Island marmot and pink rhododendron found around Rhododendron Lake. Those interested in cave exploration can visit the Horne Lake caves. Rathtrevor Provincial Park, with a sandy beach that is 1000 m wide at low tide, attracts large concentrations of waterbirds during the annual herring run in February and early March. Also flocks of brant stop for a few days from February to May.

**Birds**: white-tailed ptarmigan and gray jay.
**Amphibians:** salamanders and newts.
**Wildflowers**: dogwood (tree and dwarf), pink calypso, white trillium, white erythronium (along the river), rhododendrons, heather, penstemon, alpine saxifrage and mountain buttercup.

### Little Mountain Lookout

The viewpoint provides a magnificent picture of the Englishman River valley and Mt Arrowsmith. A walk to the east on the mountain provides a glimpse of the Strait of Georgia. Manzanita grows here, along with other spring flowers. To reach the site, drive about 3 km west of Parksville on Hwy 4 to Bellevue Rd, turn left and continue to the first paved road on the left, Little Mountain Rd. Drive to the end of the pavement and park by the microwave towers. **Caution:** The cliff on the south side of the mountain has a vertical drop of several hundred feet. The area is not fenced.

### Englishman River Falls Provincial Park

Hiking trails in this park wind through a forest of tall firs, massive cedar and hemlock. In early spring dogwoods are a mass of white blossoms. Along the mossy river bank look for white erythronium, pink calypso, white trillium, and other spring flowers. A wide variety of ferns and mosses are present. To reach the park drive west from Parksville for about 5 km to Errington Rd on the left. Follow this road an additional 8 km to the park where there are camping sites, and a covered picnic area with stoves.

### Hamilton Swamp Trail

The site contains woodland and marsh birds together with a variety of flowers. Migratory birds, including a variety of shorebirds, swans and Canada geese, stop on their way through. Roosevelt elk are found in surrounding wildlands. Plants include swamp cinquefoil, salal and oregon grape. Drive west from Parksville for 11 km; turn right at Coombs Rd, the first road after the Coombs "Old Country Market"; continue to the end of this road where it meets 4A; turn left and follow it for 0.4 km to the parking lot on the left. Follow the trail to the swamp.

### Little Qualicum Falls Provincial Park

Although very close to Englishman River Provincial Park, Little Qualicum has a very different plant community. The gravelly soils are drier, resulting in more pine in the forest. Spring flowers are great. The roadside Beaufort and Cameron Lake picnic sites are wetter than near the falls and consequently harbour different plants. Many salamanders and newts are found at these sites as they prefer the cool damp cedar and fir forest of the area. Streaks of volcanic intrusions are visible in the river bed. These rocks contrast with the older metamorphosed sedimentary layers. Located 19.2 km west of Parksville on the right side of the road.

### Mount Arrowsmith Trail

The scenery and alpine flowers are worth the all-day hike. Above the tree line watch for white-tailed ptarmigan, gray jay and other mountain birds. At the lower elevations along the trail look for spring flowers including dogwood, dwarf dogwood, and goatsbeard. At the higher elevations are heather, mountain buttercups, and alpine saxifrage. Drive west on Hwy 4 from Parksville for about 21 km to the east end of Cameron Lake. The trailhead is on the left hand side of the road. Park and climb Mt Cokely and then cross over to Mt Arrowsmith. The trail is blazed up to Mt Cokely. From Cokely to Arrowsmith is only for experienced hikers. Maps and directions are available at the Parksville Tourist Bureau.

### Cameron Lake

The woods surrounding this deep lake are full of flowering dogwood and dwarf dogwood. A private campground operates at this site. Lying 21 km west of Parksville on Hwy 4, the lake is 11 km long.

### Parksville Community Park

At low tide, the sandy beach has numerous tide pools. To enter turn left at Corfield Rd light, in downtown Parksville.

### Horne Lake Caves and Provincial Park

Four of the six undeveloped caves are open to the public. Horne Lake Main Cave, the first you encounter, has about 150 m of passages that vary from spacious chambers to small crawl spaces. Several hundred metres up river along the bluffs, Horne Lake Lower Cave has about 65 m of passages. Both of these caves have been heavily used by people with much damage as a result. The River Bend Caves with over 400 m of mapped passages have been grated to protect their fragile nature. They may be entered only by conducted tours. Contact Rathtrevor Beach Provincial Park near Parksville for a guided tour. To reach the caves drive north from Parksville for about 25.6 km on Hwy 19. Turn left at the Horne Lake Garage and continue on the Horne Lake Rd for 11 km, then turn right to the camping area another 7 km further around the lake. The caves are 1 km beyond the campsite.

### Rathtrevor Provincial Park

More than 150 species of birds have been recorded in the park. Seabirds gather for the herring spawn run February to early March. Brant are common offshore from late April to early May. The long sandy beach has numerous sand dollars and other forms of marine life. The park has an old farm house and broad fields and meadow with stands of second growth deciduous and coniferous trees, and the occasional single huge Douglas-fir. Many well marked trails wind through the area. A park naturalist is on duty every summer.

**Birds**: Hammond's flycatcher, warblers, golden-crowned kinglet, pileated woodpecker and bald eagle.

Drive south 3.2 km from Parksville on Hwy 19, and turn left (east) by the log store for another 2 km into the park.

### Rhododendron Lake

The rhododendrons are in bloom the latter part of May and early June. Kalmia, Labrador tea and other moisture loving plants also grow on the border of the lake. Only two species of rhododendron occur in the mountains -- the white *Rhododendron albiflorum*, and the pink *Rhododendron macrophyllum*. The pink species is very rare on Vancouver Island, only growing in two places -- here at Rhododendron Lake and another spot near Victoria. To get to the lake drive south from Parksville on Hwy 19 for 7.2 km and turn right into the MacMillan Bloedel Northwest Bay Camp. You **must** check at the camp office before entering. Either phone 468-7621 the day before, or go in person when you arrive to be sure the road is open.

### Green Mountain

A colony of Vancouver Island marmots reside on this mountain. These dark brown groundhoglike mammals with white muzzles are unique to the Island. Drive south to Nanaimo and then another 10 km further south on Hwy 19 and turn right (west) onto the Nanaimo Lakes Rd. Green Mtn is another approximately 55 km west from Hwy 19. You must check with the B.C. Fish and Wildlife staff in Nanaimo, phone 758-3951, before going into this area.

*Pauline Tranfield and Neil Dawe*

## MACMILLAN PROVINCIAL PARK (CATHEDRAL GROVE), PORT ALBERNI

Visit a magnificent grove of 800-year-old Douglas-firs which survived the last major forest fire of 300 years ago. Indian markings can be seen on the old trees. The trees measure three metres in diametre or slightly over nine metres in circumference. The largest tree, standing 75 metres tall, requires six adults holding hands to encircle it.

The park was donated to the B.C. parks system by the H.R. MacMillan Company in 1944.

Wildlife is scarce in MacMillan park because of the dense overstorey and lack of varied habitat. A few warblers search out insects in the crowns, but otherwise the woods are silent. Where the canopy is closed because of the wide overstorey, the undergrowth has been reduced to a few huge sword ferns and huckleberries growing through a thick carpet of mosses. The large size of the trees is due to deep soils, shelter from high winds, absence of forest fires for a long time, combined with no logging. The park lies in the glacial-carved Cameron Valley.

Other large trees to look for include Western red-cedar, western hemlock, grand fir and broadleaf maple. A self-guiding trail takes a visitor through the trees.

Further along just east of Port Alberni, on the south side of Hwy 4, look for a trail marked Roger Creek Nature Trail. This path follows Roger Creek through a second-growth fir and maple forest which can offer interesting birding.

Along with regular woodland species, it is possible to see Townsend's and black-throated gray warblers, and western and Hammond's flycatchers.

### Stamp Falls Provincial Park

North of Port Alberni, on Beaver Creek Rd, the park offers a spectacular waterfall into a canyon. Spawning salmon make use of fish ladders, with others attempting thrilling leaps up the falls.

### J.V. Clyne Sanctuary

The sanctuary was established by MacMillan Bloedel Ltd. to protect a resting and wintering area used by waterfowl, particularly trumpeter swans. This undeveloped sanctuary offers good birding opportunities especially during spring and fall migration.

**Birds:** shorebirds, Lincoln's sparrow, warblers, waterfowl, northern shrike, tundra swan, Eurasian wigeon, bald eagle, northern harrier, merlin, marsh wren, and black and Vaux's swifts.

The area is also of interest for plants. Recent discoveries indicate some plant species present here were found previously only as far south as Alaska.

**Further Information**

Travelling west from the Island Hwy at Parksville take Hwy 4. Camping is available at Englishman River Falls and Little Qualicum Falls Provincial Parks. Both of these sites have good trails, beautiful cascades and swimming holes. MacMillan Provincial Park lies 32 km west of Parksville on the west end of Cameron Lake, and 16 km east of Port Alberni along Hwy 4, west of Parksville. From Port Alberni the J.V. Clyne Sanctuary is reached by driving 2.7 km west of Clutesi Haven Marina, cross the steel bridge over the Somas River and turn left at the sign to the airport. A further 0.8 km where the road forks, veer to the right as the left fork goes to the airport. Drive another 1.6 km and watch for the sanctuary sign on the left. There are two parking areas along the road into the sanctuary. The Visitor Centre in Port Alberni sits on the highway on the way into town, and has an excellent summary of trails together with up-to-date fire information.

*David Stirling and Bruce Whittington*

## PACIFIC RIM NATIONAL PARK

Situated on the west coast of Vancouver Island, the park is a thin strip of land and islands fronting on the Pacific Ocean. Both marine and upland habitats are present including deep forest, bog and muskeg. Watch for migrating whales in spring and fall at Long Beach, the most accessible and northern section, or take a tour out of nearby Tofino or Ucluelet. The beach is also one of the few easily accessible stretches of exposed rocky and sandy shores on the Island. Lying west of Parksville, the park is about a five-hour drive from Victoria.

The park contains Long Beach with two major beaches, the Broken Islands Group with about a hundred islands and rocks, and the West Coast Trail. This hiking path, 73 km in length, includes the historic Lifesaving Trail which was built for shipwrecked sailors who made it to shore after a shipwreck on this "Graveyard of the Pacific." Biologists have in the past recorded nearly 260 species of birds with close to 60 breeding residents.

Islands off Long Beach host breeding populations of pelagic cormorants, the only known Canadian colony of Brandt's cormorant, black oystercatcher, pigeon guillemot and glaucous-winged gull.

Rocky intertidal sea creatures are common at Green Point and Box Island on the north end of the beach. Chesterman Beach and Frank Island, north of the park boundary, there are excellent spots to see intertidal life. Occasionally "purple sailors," a jelly fish called *Velella*, are washed up on the beaches in summer. Every five or six years there is an influx of mole crabs (*Emerita*) on the sandy beaches.

Many more colorful undersea animals live in deeper waters. About three times each summer, park divers collect all kinds of deep-water plants and animals and place them in shallow pools for a short time. Park staff lead tours to these spots. After the tours, divers return the creatures to the sea. Phone ahead to coordinate your trip with one of these "Scuba Specials."

Wooded bogs are present in lowlands behind the beaches where fresh water accumulates from the approximately 300 cm of rain each year. Impervious clay layers under the soil trap the water on the surface. In bogs, sphagnum moss decays under-water, resulting in highly acidic conditions. Labrador tea and bog laurel also thrive in the bog. Few other plants can survive in this acidic environment, but shorepine, here a stunted and clump-topped tree with a distinctive silhouette, grows in these wet areas. In other parts of British Columbia this tree grows tall and straight and is known as lodgepole pine. Around the margin of the bogs, evergreen huckleberry grows in profusion.

In the spring Scotch broom provides a dazzling display of yellow along the roadsides. Introduced to Vancouver Island a little over 100 years ago, it has spread rapidly.

Forestry plantations cover the recently logged parklands. Here fireweed hides the charred snags and stumps. Crews have planted Douglas-fir, amabilis fir and Sitka spruce. Western red-cedar and hemlock have come up on their own.

The sub-tidal zone off the park provides ideal habitat for kelp which sometimes grows as deep as 30 metres. Animals living on the sand bottom adjacent to this vegetation attract gray whales as they migrate through the area.

Since this park lies on the edge of the continent, the rocks found here have been modified greatly through plate tectonics that ripped, tore, melted and compressed them. Box Island, at the north end of Long Beach shows mashed and tilted layers of argilite (originally shales which came from clays) and ribbon chert (from sandstone derived from beach sands). Radar Hill, nearby, is made up of greywacke (also originating from shale and sandstones).

**Birds:** red-throated loon, Brandt's and pelagic cormorant, great blue heron, bald eagle, black oystercatcher, glaucous-winged gull, Steller's jay, hermit thrush, Townsend's

warbler, pileated woodpecker, chestnut-backed chickadee, golden-crowned kinglet, winter wren, brown creeper, red crossbill, fork-tailed and Leach's storm-petrel, marbled murrelet, rhinoceros auklet, tufted puffin, band-tailed pigeon, rufous hummingbird, Swainson's thrush, orange-crowned warbler, four species of shearwater and Caspian tern.
**Sea mammals**: northern sea lion and gray whale (late March and early April)
**Plants:** Salal, Pacific crabapple, black twinberry, Western Red-Cedar and Western Hemlock (many cedar over 200 years old), red huckleberry and blueberry shrubs, grand fir, (in ideal locations this tree becomes the tallest tree in the park), Western yew, false azalea, red alder, several species of willow, cascara, salmonberry and thimbleberry.

## Wya Point

A good spot for to observe intertidal life, it lies at the south end of Florencia Bay below Long Beach. Four trails lead to the Bay, with a nice walk to Wya Point. Willowbrae Trail, the southernmost one, starts outside the park and emerges near the point. This half hour stroll passes through the forest, narrow foreshore and then into the intertidal area.

## West Coast Trail

The hiking trail provides an excellent opportunity to experience the west coast at its best. Numerous booklets and pamphlets have been written on this trail (see below), and can be obtained from the Royal B.C. Museum Gift shop or most bookstores in the province. During the summer, you must make reservations to hike this trail.

## Schooner Cove Trail

The trail leads from a parking lot 14 km south of Tofino. This winding trail (0.8 km) descends through cedar and hemlock forest, and finally through a Sitka spruce fringe. The relatively open tent sites are surrounded by salal bushes.

## Green Point

Near the top of the cliffs the tallest cedars show effects of the strong sea breezes. Salt spray will eventually kill them. These trees have developed heavy side branches that are almost secondary and tertiary trunks, which extend outward and then upward. At the bottom of the hill, near the pit toilets one branch of the trail runs behind the misshapen Sitka spruce adjacent to the foreshore. The other short branch leads to the beach.

## Rain Forest Trail

Along this trail amabilis fir grows along with cedar and hemlock in this fully mature climax rain forest. The trail contains two short loops, one on either side of Hwy 4 about 6 km north of the information centre.

**Further Information**

To reach the park drive north on Hwy 1 from Victoria to Parksville, and then west on Hwy 4 to the entrance. For a good overview of the park, a visit to Wickaninnish Centre, the interpretive centre for the park, is a must. Two campgrounds occur in the park: the main one at Green Point and a walk-in site at Schooner. Primitive camping is provided on some of the islands in the Broken Group Islands including Clark, Gilbert, Hand, Willis, Benson, Turret and Gibraltar. Wilderness camping is done along the West Coast Trail.

Those interested in trips to see whales can go to Tofino or Ucluelet where whale tour boats may be arranged in spring and fall. • Recommended reading: *The West Coast Trail and Nitinat Lakes*, published by Douglas and McIntyre, 1978, a Sierra Club book; *Birds of Pacific Rim National Park*, Occasional Paper No. 20, by the B.C. Provincial Museum; *Canada's National Parks, A Visitors Guide* by Marylee Stephenson, published by Prentice-Hall Canada. The park has produced a variety of materials including a bird check list, park brochure, and pamphlets on hiking which are available free from the information centre on Hwy 4, or by writing the park at Box 280, Ucluelet, B.C. V0R 3A0. Telephone them at (604) 726-7721.

Canada's National Parks *by Marylee Stephenson*
*Phil Lambert, Royal B.C. Museum*

# CLAYOQUOT SOUND AND MAQUINNA PROVINCIAL PARK

Gray whales migrate through here with peak numbers in late March and early April. About 50 whales remain all summer providing excellent whale watching opportunities out of Tofino. Many day tours are available. Meares Island has the largest cedar trees left on the B.C. coast. Rich shellfish beds are found off its shores. Flores and Vargas Islands contain old growth forests and miles of wilderness and beaches. A puffin colony is found on Clelland Island. Visit the excellent hot springs in Maquinna Provincial Park. The springs are natural, undeveloped and with water temperatures in excess of 50° C. There are natural showers and pools, too. The boardwalk from the government wharf in Hot Springs Cove passes through lush coastal rain forest. The protected route from Tofino to Hot Springs Cove via Miller Channel and Shelter Inlet is suitable for small craft. The outside waters are exposed but good on calm days. Expect fog by late summer. A new hiking trail has been completed from the head of Bedwell Sound to near Buttle Lake in Strathcona Park. The quiet and rich waters along the sound provide a haven for a wide selection of birdlife.

**Birds**: Albatross (offshore), shearwater, osprey, bald eagle, sea ducks, alcids, cormorant.
**Marine Mammals:** Gray whales(April through November), orca, sea lion, mink, river otter and harbour seal.

**Further Information**
Tofino, the closest community, lies at the north end of Hwy 4. From here either fly into Hot Springs Cove or go by boat. Tofino has boat ramps and many boat and plane charters. There are government wharves at Ahousat and Hot Springs Cove. Maquinna Park lies on a peninsula north of Flores Island. • Hot Springs Cove has no developed campsites. For whale watching, sailing, nature photography and hot springs tours contact The Whale Centre, Box 394, Tofino, V0R 2Z0; or Tofino Adventures Booking Centre, Box 620, Tofino, V0R 2Z0, (604) 725-4220. Tours in sea kayaks and native dugouts are also available. • The Hesquiaht Band offers accommodations at Hot Springs Cove Lodge. • Marine charts to use include 3649 and 3643.

*Bob Sutherland*

# COURTENAY - COMOX AND FORBIDDEN PLATEAU

Concentrations of trumpeter swans, ducks, loons, grebes, cormorants, gulls and shorebirds are found from fall through early spring in the Comox valley. During spawning season many bald eagles appear. Mild annual temperatures and generally open winters hold a variety of birds all year. The Courtenay River Estuary is considered by some to be one of the finest estuaries in Canada.

## Hamilton Mack Laing Nature Park

The site contains a small salmon stream, coastal forest, and beach with easy trails for exploration. The park is an excellent and easily accessible location to watch the rich bird life of Comox Bay. Winter is the best time. Snow geese sometimes appear. Except for high tide, you can walk from here southeast, to the left, along Croteau Beach and around the inner bay of Goose Spit. The park lies about 500 m east of Filberg Lodge and Gardens, both owned by the town of Comox.

## Courtenay River Estuary

This waterfowl wintering habitat is favoured by hundreds of trumpeter swans. Over 145 species of birds may be observed in the course of a year. In addition, the green belt adjoining the sawmill has produced 84 species of plants, including wild hollyhock in late spring.

To explore the area take an eight km drive along Comox Rd and along the south side of the Courtenay River. Begin at the foot of Comox Hill on Comox Ave on the west side of Comox. Search here for shorebirds. Drive west to the observation point just beyond the Cement Tower to find bay birds. Across the highway, pond and bay ducks and shorebirds can sometimes be found at the slough. In summer there are many land birds, including turkey vultures. In spring, wildflowers are found here and near the Sawmill to the west. Continue west, and then cross 17th Street bridge and turn left, or south, along Cliff Ave to reach Mansfield Drive. This is a good spot for pond and bay ducks, and hawks. Continue down the Island Hwy and turn left at Millards Creek to see a variety of bay and sea ducks. Move on south to Royston and turn left on to Marine Drive to find more bay and pond waterfowl, and many seals.

## Puntledge Park

The site contains numerous flowers in spring, and hundreds of coho and chum salmon spawn in fall. White and pink Easter lilies are found, together with chocolate lilies.

**Birds:** bald eagles, little green heron and red-breasted sapsuckers.
**Plants:**The numerous footpaths pass through Douglas-fir, maples and a good variety of shrubs.

To reach the site which sits about 10 blocks west of downtown Courtenay, take 5th Ave to Menzies and follow it to the Park.

## Point Holms

Approximately one km of open shoreline is available to the public all year. Sea bird watching is excellent from here. The beach provides a good opportunity to find sea animals including oysters and clams.

The low tides in May, June and July bring many forms of intertidal life into easy access for the walker. Look for sea cucumbers, sea urchins, sea stars, nudibranches, sea worms, starfish and oysters.

**Birds:** (winter) red-breasted mergansers, scaups, scoters, loons and harlequin ducks.

Access is east along Lazo Rd approximately 4.5 km from the mall in Comox. There is a wildlife sanctuary on both sides of Lazo Rd on the way to the point. An osprey nest lies closer to the point also on the left.

## Kye Bay Beach

At low tides there are huge expanses of sand which contain numerous live sand dollars and moon snails. Summer is best for sea life an winter for sea birds. In mid-May whimbrels are seen here, especially in the north-western area.

At low tide take the approximately 4-km stroll along Kye Bay southeast around to Point Holmes. A further extension of the walk would be another 3.5 km south along the pebble and boulder beach to the sand beach below the massive Willemar Bluffs to Goose Spit Regional Beach. To reach this site 5.5 km from Comox, take Balmoral Ave and Lazo Rd past Point Holmes, and turn right on to Kye Bay Rd. Contact the Tourist Bureau on Cliffe Ave (the highway) in Courtenay for information.

## Seal Bay Regional Nature Park

The 24 km of easy trails allow the young and elderly to explore a 1.2 km natural rocky shoreline with some sandy areas. The site is of interest in all seasons, with swamps, dry gravel places and heavily treed to light bush. Large ravines lead to shoreline. One hundred and fifteen species of land and water birds have been seen here, and sixteen species of mammals including seals. Fungi appear in abundance in the fall. A wide collection of ferns and flowering plants occur, too.

To reach the park take Anderton Rd north from Comox to Waveland and west to Bates Rd which runs through the site. Trails branch off along either side of Bates. Brochures are available at the Tourist Bureau.

## Cumberland Sewage Lagoon

Cumberland, southwest of and higher than Courtenay, is an old mining town where, over the years, eight mines were developed. The nearby lagoons provide good habitat for green-winged teal, ring-necked duck, shoveler and other water birds including Canada geese. Marsh wrens use the cattail area to raise young each year. MacGillivray's warbler, yellowthroat and varied thrush are here in spring and summer together with red-tailed hawk. Northern shrike has been found in winter. Beaver and muskrat are found all year.

### West of Cumberland

Beyond old "Chinatown" is a marsh and pond flowing into Coal Creek. Look for Canada geese, sapsuckers, marsh wrens, and possibly a rail.

### Comox Lake Bluffs

These high south-facing bluffs above the lake provide unique habitat for a wide range of flowers which are best in April-May. This is the northern limits of arbutus and hairy manzanita. Kinnickinnick predominates here and hybridizes with manzanita. Ferns and calypso orchids are also present in abundance. To reach the site take Lake Trail Rd west from Courtenay, to Comox Lake. Proceed about 0.8 km beyond Bevan Gate on the Crown Forest road and pull off to park on the left-hand side by the steep access road.

### Denman and Hornby Islands

These two islands lie 19 km south of Courtenay. They are a 10 minute ferry ride apart. Killer whales are sometimes seen during a summer passage. The islands are somewhat warmer and drier than the adjacent Vancouver Island. As a result Garry oak, arbutus, and tall mahonia thrive at their northern limit. Mockorange also grows here.

On Denman, the 80-acre Fillongley Park affords pleasant picnicking and camping. On the south end of the Island, just before the ferry slip to Hornby, take the road to your right. It leads to the trail on the high bluffs of Boyle Point overlooking Chrome Island lighthouse. Here, in March, sea lions may be seen on the rocks and in the water during the herring run. Many eagles and herons are here all year. Sea birds abound in fall, winter and spring. Metcalf Bay, near the southwest end of Denman, is a wintering ground for thousands of loons, grebes, cormorants, ducks, and gulls.

Helliwell Provincial Park, on Hornby Island, has an exceptional 3-km circle walk to Helliwell Bluffs. Major groves of Garry oak lie on the southwest corner of the park. Spring flowers are found in profusion on the dry bluffs in April. The shooting stars are especially beautiful at that time and into May. To view the petroglyphs take the road to Tralee Point. At low tide one can walk along the beach to these rock paintings.

### Tree or Sandy Island Provincial Park

The park contains Tree Island plus the three Seal Islets. At low tide you can walk to the point of Denman Island and back on the connecting sandy area. In spring and fall the site is good for migrating birds. It also contains a good collection of Gulf Island plants and beautiful spring flowers. Many shells are found on the sandy beaches. To reach the site use a private or charted boat.

### Paradise Meadows

The meadows, part of Strathcona Provincial Park, are a sectacular subalpine area for flowers and are best in mid-July to early August. Fall colours are great in September. There is an easy short trail down into the meadows from the Mt Washington Ski Rd and an excellent trail circles the meadow. To reach the site from Courtenay (directional signs are easy to follow.) take Island Hwy north, turn left on Headquarters Rd, Piercy Rd to Duncan Main logging road. Turn right onto this logging road and follow the Mt Washington Ski Hill signs. Contact the Tourist Bureau for further information.

### Forbidden Plateau

This part of Strathcona Provincial Park has gray jay, raven, junco, hermit and varied thrush, and flycatchers. Ptarmigan and blue grouse are present, too. This is a delightful area in which to hike in subalpine vegetation, including shooting stars (mid-July), and lilies in the ponds. Vegetation includes yellow cedar, amabilis fir, mountain hemlock, slide alder and heathers. Other plants include Trolius, blueberry, copper bush, rhododendron and buck bean. The snow remains until mid-July with the season open until about October 15. For access from Paradise Meadows follow the well-marked hiking trails on to the Plateau. A hike to Battleship Lake - Helen McKenzie Lakes makes a delightful nature wander. On day hikes do not go more than 4 or 5 km into the park and be prepared for bad weather at any time. Contact the Tourist Information Bureau for further help.

### Woodhus Slough

This body of fresh water, lying very near Georgia Strait, is managed by the B.C. Wildlife Branch. Formerly a salt water slough, it was closed off to the sea and has become a sanctuary for waterfowl. The site is a prime feeding and resting area for migrating birds, especially ducks and geese. Trumpeter swans frequent the water and fields in winter. Bald eagles are common. The site is also rich in plants including lichens, mosses, many shrub species and wild onions. In spring look for masses of blue-eyed Mary and goldstar.

To reach this spot drive about 27 km north of Courtenay to about 3 km north of the Oyster River Bridge. Turn right on Salmon Pt Rd and proceed about 1 km to the end of the road.

**Further Information**

*Hiking trails III: Central and Northern Vancouver Island.* Outdoor Club of Victoria. 6th ed. 1986. *Trails and Walks in the Comox Valley.* F.M. Johnson, Lindsay Press, 1988. • For more information contact the Comox-Strathcona Natural History Society, Box 3222, Courtenay, B.C. V9N SN4.

*Members of the Comox-Strathcona Natural History Society*
*coordinated by Fran Johnson*

# MIRACLE BEACH PROVINCIAL PARK AND MITLENATCH ISLAND PROVINCIAL NATURE PARK

Miracle Beach Park, with 334 acres, contains some virgin timber, a coho salmon spawning run up Black Creek (Oct.-Nov.), and an excellent swimming beach with lots of marine life. Birding is best in spring and fall when many woodland and sea birds are around. Black Creek estuary, on the north boundary, is a good spot in all seasons.

Mitlenatch Island, an 88-acre reserve, has early spring flowers, nesting seabirds, abundant marine life and good snorkeling and scuba diving. The Island lies in a rain shadow and is considerably drier than neighboring land masses. It also has numerous garter snakes that are larger than average. Over 150 bird species have been recorded for Mitlenatch. It contains large nesting colonies of glaucous-winged gulls and pelagic cormorants .

**Birds:** (Miracle Beach) bald eagle, varied thrush, black swift, seabirds and numerous brant each spring; (Mitlenatch) pigeon guillemot,black oystercatcher, glaucous-winged gull, pelagic cormorant.
**Mammals**: raccoon, deer, occasional black bear and wolf.
**Marine mammals**: Killer whales, occasional gray whale, sea lion, river otter.
**Marine life:** sand dollars, moon snails, horse clams, seaweeds.
**Plants:** (Black Creek) erythronium lilies (April); (Mitlenatch Island) prickly pear cactus, mature Douglas-fir, tiger lilies, blue camas, white hyacinths, hare bells, shorepine (lodgepole pine).

**Further Information**
Miracle Beach lies 24 km north of Courtenay. Reach Mitlenatch Island by boat from Salmon Point on Miracle Beach or Campbell River. There is no boat dock on Mitlenatch, but you will find a tiny harbour with two anchorages. Pamphlets for both sites are available at the Nature House in Miracle Beach.

# CAMPBELL RIVER AND QUADRA ISLAND

From Campbell River it is easy to explore marine habitats with scuba gear or hike rocky shorelines, salmon streams, and woodlands for birds and flowers. A total of 265 species of birds with 83 recorded as nesting, is given on the area's bird checklist available from the Mitlenatch Field Naturalist's Society, Box 413, Heriot Bay, B.C. V0P 1H0.

## ▶ Willow Point Reef

Discovery Passage has very strong tidal currents which support an abundance of marine life, and Willow Point Reef, which is mainly boulders, gravel and rock, is one of the few places to gain easy access to

Black Oystercatcher, permanent resident on the saltwater outer coast and rocky islands of B.C. Nests in depression of bare rock or hollow in beach gravel.

Sea palm, *(Postelsia palmiformis)*, occurs only where the surf hits hardest. Instead of withstanding the wave shock with a rigid stalk, this brown alga has a tough but flexible stalk that bends with the forces to dissipate the high energy and rebound to an upright position.

**Page before:** Kaisan village site, west coast Queen Charlotte Islands.

Northern or Steller's sea lion is a continuous feeder except during the breeding season. One study showed diet was squid and octopus 36%, clams 29%, sand lance 25%, rock fish 11%, crabs 9%, flounder and greenling each 4%, halibut and lumpfish each 2%, and 18% unidentified fish remains.

Desolation Sound.

Red rock crab, *(Cancer productus)*, buries itself in sand or gravel under rocks during low tide. When the tide comes in, the crab stalks its prey of snails and other shellfish.

Leather star, *(Dermasterias imbricata)*, inhabits rocky shores in sheltered and exposed waters. In sheltered areas sea cucumbers make up most of its diet. In exposed sites this starfish is sub tidal and feeds on anemones.

Giant green anemone, *(Anthopleura xanthogrammica)*, inhabits cracks and crevices, on the most exposed parts of the coast, waiting to grab small animals dislodged by the waves.

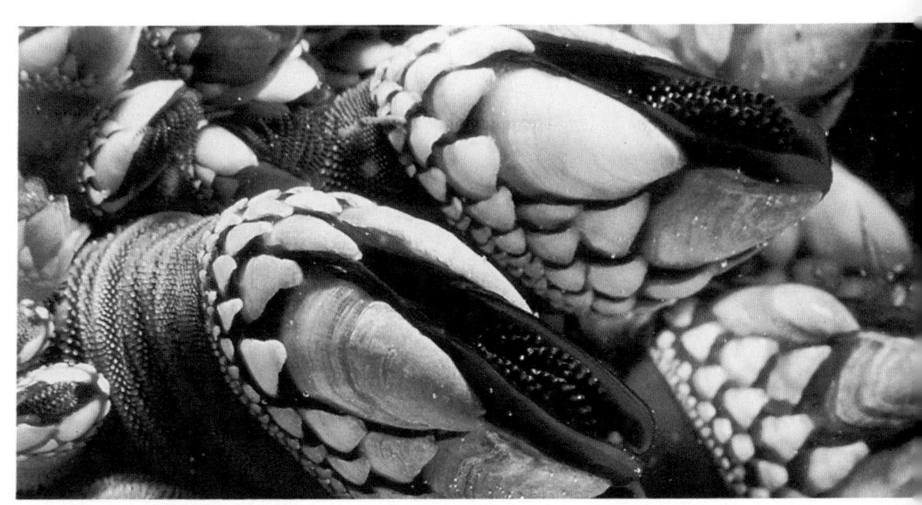

Goose barnacle, *(Pollicipes polymerus)*, is part of the community of animals in the upper two thirds of the intertidal community along the exposed coast of B.C.

**Next page:** Whitesided or Pacific striped dolphin found in the sub tropical waters of the north Pacific. In winter it occurs regularly in protected seas at the north and south ends of Vancouver Island.

Dense rookeries of northern or Steller's sea lions occur in the breeding season. Mature bulls, about double the size of the cows, fiercely defend harems for this short season. The rest of the year they swim and feed in herds and haul up onto rocky islets on calm, sunny afternoons to rest. They stay in the sea during stormy weather.

Evening at Tofino.

Hairtrigger Lake, Strathcona Provincial Park.

Blue grouse females in breeding season inhabit lowland clearcut and burned areas, open mountainsides, meadows, forest edges and openings, and high up near the treeline. After the breeding season and in winter they inhabit coniferous forests at higher altitudes.

Poplars in the fall.

Black banded skimmer, *(Libello forensis)*, a memberof the skimming dragonfilies which fly low over ponds taking mosquitos and midges as major prey.

Mule deer are gregarious, often in small herds of single sexes in summer but gathering in larger groups of all ages and sexes in winter. An older experienced doe acts as leader.

White bog orchid (*Habenaria dilatata*), horsetail *(Equisitium* sp.*)*, paintbrush (*Castilleja* sp.*)*, and introduced timothy *(Phleum* sp.*)*.

The snowshoe hare has a population cycle of about ten years. Its major predators also cycle in numbers, usually peaking and crashing a year later than the hares.

Grizzly bear in proposed Khutzeymateen Grizzly Sanctuary north of Prince Rupert, B.C.

The name Penstemon comes from *pente* (five) and *stemon* (thread) since the flower has five stamen or male parts, one sterile and four fertile.

Mountain goats are moderately gregarious, forming small herds of nannies and offspring in summer. Billies remain alone or with other billies but join the nanny groups in breeding season in autumn. Social order is matriarchal, with nannies dominating the more peaceful billies. Goats spend much time bedded down on mountain slopes , cliffs and ledges with broad views of the surroundings, where they doze and chew cud.

Pine Pass is in the northern Rocky Mountains. Here several eastern and northern plants and birds species meet and a few southern B.C. species probably reach their northern limits. Boreal birds such as three-toed woodpeckers, spruce grouse and northern hawk-owl occur year-round.

Alpine flower meadows with paintbrush, anenome *(Anemone occidentalis)*, Erigeron, lupine and Senecio, in the proposed White Grizzly Wilderness Area.

Butterflies, like the swallowtail, sip nectar from flowers such as dandelions.

Killdeers favour open interior uplands and are among the earliest birds to appear in spring, soon after the first patches of bare ground show through the snow.

Forest floor with polypore bracket fungus and bunchberry leaves.

**Next page:** Winter in central B.C.

the water without scuba gear. Go at low tide and don't get caught out at the end of the reef during a rapidly rising tide, especially in storm conditions.

Waterfowl overwinter here in abundance, including loons and sea ducks.

**Birds:** oystercatcher, great blue heron; shore birds and brant are seasonal.
**Marine Life:**seaweed, nudibranchs, sea stars, anemones, sea cucumbers, and small crabs and snails

The reef lies about 5 km south of downtown Campbell River, and about 0.25 km north of the road to the airport. On the east side of the highway look for picnic tables and a three-car parking lot with beach access. The reef is visible only at low tide. Tide charts are available free at sports shops. The municipal bus serves the site. For bus times phone 287-7433.

## Haig-Brown / Kingfisher Creek Heritage property

This site was the home of Roderick Haig-Brown, a very influential B.C. conservationist and a famous writer of outdoor books on fishing, conservation, rivers and wild country. His wish to have Kingfisher Creek returned to its original stream bed so that the salmon would again spawn there (a man-made outflow confused the fish), had been fulfilled as of 1984. The Haig-Brown barn has been renovated to serve as a meeting/teaching area, and boardwalks hav been built beside the spawning channel through the low-lying alder bush which is a favorite spot for small birds. In addition, there is a self-guided ethnobotanical walk which features plantings of native shrubs and trees which were used by the Indians.

This provincially owned site lies along Hwy 28, the Gold River Rd, opposite the junction to the fish hatchery, 3 km west of downtown Campbell River. The Quinsam provincial campsite in Elk Falls Park lies just one km west.

## Quisam Nature Trails

This site contains the federal Quisam Salmon Hatchery, provincial campsite, and a 4-km one-way walking trail which is a beautiful walk along the river. Watch for kingfisher, great blue heron, and bald eagle when the salmon are spawning. Deer occur and small mammals may also be seen. From late August to November spawning salmon include chinook, pink, chum and some sockeye. Steelhead spawn in winter. This is one of the largest hatcheries in B.C.

The site lies 3.9 km west of downown Campbell River on Hwy 28. Park at the campsite or the fish hatchery. Trail brochures are available at the hatchery.

## Ripple Rock Trail

The 4.7-km trail leads from the parking lot to Seymour Narrows. Ripple Rock, in the middle of these narrows, was a major hazard to navigation along the inside passage, and was the cause of many shipwrecks. Much of the rock was blasted away in 1958, and the detonation was, at that time, the world's largest non-nuclear explosion.

The trail winds between the bays and forests, and there are many good viewpoints along the bluffs as you hike in.

**Birds:** sea ducks.
**Marine Mammals:** harbour seals.
**Trees:** Douglas-fir forest.
**Wildflowers:** (April and May) avalanche lily, pink fawn lily, camas, chocolate lily, wild onion and trillium.

To reach the site, drive north 18.6 km from downtown Campbell River on Hwy 19, the road to Port Hardy. Pass the Bloedell Pulp Mill and watch for the hydro wires crossing the road. Park on the east side of the road in the small lot and from here begin the hike to the east. The public bus from Campbell River to Port Hardy passes the site, and you can arrange to be dropped off. Contact Marcie Wolter who helped to build this trail. Phone her at 285-3849.

## Quadra Island Ferry Ride

The ride provides easy access to Discovery Passage which is a very rich marine environment due to the strong tidal currents. This passage is a very productive sports salmon fishing area. It also offers superb scuba dive sites. Use caution as the tidal currents are dangerous. Obtain local advice.

**Birds:** (winter) common murre, marbled murrelet, rhinocerous auklet, loons, grebes, cormorants, diving and sea ducks, great blue heron, bald eagle and gulls.
**Marine Mammals:** killer whles, Dall's porpoise, harbour porpoise and sea lions.

## Quadra Island - Cape Mudge

High sand bluffs, a low-tide reef, petroglyphs, an ancient Indian village site and midden, large grand firs, a lighthouse, and a superb view await the visitor to this site. There is also a good reef with tide pools, rocks and boulders.

**Birds:** bald eagle, great blue heron, sea ducks, gulls and shorebirds
**Marine life:** sea stars, nudibranchs and small crabs.

To reach the Cape, head south from the ferry landing on Green Rd. Stop at Cape Mudge Village and the Kwaqiulth Museum. hen drive east a short distance on We Way Rd to Joyce Rd, and then south to Lighthouse Rd. Take it to the end and turn south a short distance to the lighthouse and park. Hike the approximately one-km trail past the Tsa-Kwa-Luten Lodge to the bluffs and the site of the Old Salish village that lie to the south.

## Quadra Island - Rebecca Spit Provincial Park

This 2-km-long spit of land, only 20 m wide in places, is an excellent spot for surveying sea birds, particularly in winter. Herring spawn in March along the beaches and attract thousands of ducks and gulls.

Walking and cycling is easy along the spit, and large Douglas-firs grow along there with an understory of ocean spray, snowberry, sword fern, salal and numerous wildflowers.

**Birds:** loons, grebes, mergansers, harlequins, scaup, goldeneyes, old squaws and scoters.
**Marine Mammals:** (occasionally seen) killer whale, harbour porpoise, and sea lion.

To reach the park from the ferry dock in Quathiaski Cove, drive east to the centre of the island and then 5 km north on Heriot Bay Rd. A private campground lies at the entrance, and another camp ground is found nearby at Heriot Bay.

**Further Information**
Quadra Island sits just east of Campbell River, and the ferry (a 15-minute ride) runs every hour (on the half hour from Campbell River, and on the hour from Quadra). There are no buses on Quadra, but taxi service is available at 285-3598. Camping is available just south of Rebecca Spit Provincial Park and at Heriot Bay Inn. There is gas at Quathiaski Cove, but it is closed on Sundays. A list of trails in the Campbell River Area is available from the Chamber of Commerce at the local Tourist Bureau. Phone 286-0764. Look for information about Quadra trails at Island stores. Some trails are also described in Elaine Jones' book *Northern Gulf Islands*, 1991, published by Whitecap Books.

*Bob Sutherland*

# GOLD RIVER AND WEST SIDE STRATHCONA PROVINCIAL PARK

Hwy 28, the road to Gold River, cuts across Vancouver Island from Campbell River on the east to Gold River on the west. It is a beautiful drive and wildlife is commonly seen in the early morning or late afternoon. This road also provides access to the west side of Strathcona Provincial Park, the first provincial park to be established in B.C.

## Boulding Bog and Swan Bay

On driving in from Campbell River, just after passing Strathcona Lodge, one can stop at a typical acid sphagnum bog of about 100 acres with a board walk. From here a walk can be made to Swan Bay on Upper Campbell Lake which is a wintering site for trumpeter swans from mid-November to March. The bog plants are in bloom from late May through early July.

**Birds**: blue and ruffed grouse, common raven, olive-sided and western flycatcher, yellowthroat, orange-crowned, Townsend's, Wlson's, yellow warbler, Canada goose and common merganser, trumpeter swan in winter.
**Mammals:** wolf, cougar, black-tailed deer and black bear; beaver lodges.
**Plants:** sundew, Labrador tea, kalmia, andromeda, bog cranberry, cloudberry, false asphodel, sweet gale, cascara, water lily, sphagnum moss; lady fern, maidenhair fern, wild ginger, Indian hellebore and yellow violet.

To reach the site, drive about 35 km west of Campbell River to Strathcona Lodge. Continue about a kilometre further to a gravel pit on the west side of the road. Park here and walk east across the road to the trail. Take this trail east for about 200 metres and turn right through alders to the open area and bog. After examining the bog, retrace your steps to the trail and continue onward in a loop around the bog and back to the road, approximately one km south of where you parked. Cross the road and follow the trail to Swan Bay. Address for the Lodge is Box 2160, Campbell River B.C. V9W 5C9; or phone 286-3122.

## West Side of Strathcona Provincial Park

This nearly one-half-million-acre park is a vast and magnificent area of mountains and valleys containing virgin rain forest, subalpine forest, and alpine habitat. It also contains Buttle Lake, a 30-km-long reservoir, many smaller mountain lakes, rivers, glaciers and permanent snow fields. Trumpeter swans frequently winter at the extreme south end of Buttle Lake.

Established in 1911, this is the oldest park in the B.C. system and is home to the highest peaks found on Vancouver Island, including Mt Elkhorn and the Golden Hinde. The rugged off-road backcountry areas are seldom visited even in the summer months, except along the few marked trails. The rocks are very jumbled up with mostly old volcanics, basalts, granites and rhyolite. Some limestone/ marble with caves is found in Marble Meadows.

**Mammals**: black bear, wolf, cougar, Roosevelt elk, black-tailed deer, marten, beaver, raccoon, red squirrel, ermine, and mink.
**Trees:** Douglas-fir, western hemlock, red-cedar, some amabilis and grand firs, western white pine and lodgepole pine (rain forest); yellow cedar, mountain hemlock, and alpine fir (subalpine).
**Fish**: cutthroat, rainbow, and Dolly varden.

This provincial park may be most easily accessed on the east side near Courtenay at Forbidden Plateau, Mt Washington or west of Campbell River near Strathcona Lodge. Park Headquarters lie a 1/2 km south of the Buttle Lake bridge, a few km south of Strathcona Lodge. A bus runs to and from Campbell River and Gold River on Sundays, Tuesdays and Thursdays.

## Lupine Trail in Strathcona Park

This short 2-km loop trail leads through some magnificent, old growth rain forest to Lupine Falls. The trail is found south of the park headquarters and is clearly marked by a sign on the east side of the road.

## Marble Meadows in Strathcona Park

The meadows contain beautiful alpine flora, and numerous and varied fossils are found in the limestone beds. Marsh Marigold Lake has been stocked with Kamloops trout. The area is usually free of snow by mid-August and stay lovely until mid-October. This is one of the choice sites in the region, but it is a long hike and in places the rocky terrain makes for difficult walking. Proper gear is required, and be prepared to get weathered in.

**Plants**: paintbrush, cow parsip, arnica, marsh marigold, globe flowers, penstemon, phlox, mimulus, hellebore, white rhododendron, and copper bush.
**Trees:** Firs, maples, dogwood, mountain hemlock, yellow cedar and alpine fir.

To access this site, use a canoe or rowboat. Drive to Campbell River, and then west on Hwy 28 past Strathcona Lodge. At the road junction on the east side of Buttle Lake bridge, take the south leg for over 20 km to immediately across from Phillips Creek at the Augerpoint rest stop. Put your boat in the lake and cross to the mouth of the creek. The 11-km trail west to the alpine meadows begins here. Make sure to carry overnight camping equipment along. For further information contact Ruth Masters at 334-2270.

## Flower Ridge Trail in Strathcona Park

This is the easiest route to the alpine habitat on the west side of the park. However it is steep and eroded in places and is a long day hike, with most people camping overnight at the top. The trail is about a 1220 metre gain in a3-km walk. A gentle rolling ridge extends for approximately 8 km at around 1500 metres elevation. It is worth the walk to see the alpine flowers which are best in late July and August.

**Birds:** water pipit, rosy finch, white-tailed ptarmigan and gray jay.
**Plants:** moss heather, pink and yellow heather, Menzies' penstemon, saxifrages, grouse foot, false valerian, elephant head, and spreading phlox.

The trailhead lies 28 km south of Buttle Lake Bridge, approximately four km south of the Ralph River Campground, just south of Henshaw Creek, on the east side of the road.

## Elk River Trail and Valley in Strathcona Park

This is a delightful valley with a variety of mushrooms and other fungi in late spring and fall. In summer watch for harlequin ducks in the upper rapids where they have been known to breed. Roosevelt elk are often present in winter along the highway, the power lines, and the flats of the Lower Elk Valley.

Most of the hike in passes through climax forest. Gravel flats act as a "sun-trap" providing habitat for a variety of wild flowers including arnica, mountain ash, maidenhair ferns, and pink and yellow mimulus. Flowers begin blooming in June and peak in late July/August.

**Birds**: winter wren, varied thrush, Steller's jay, raven, chickadees, kinglets, dipper, and blue grouse.
**Mammals:** elk, black bear, deer, and even wolf and cougar.

The entire hike from the parking lot to Landslide Lake, at the foot of Mt Colonel Foster, is a fairly strenuous walk of 11 km; however, it is the easiest long hike in the park.

To reach the trail head and Elk Valley, drive 15 km west of the Buttle Lake bridge on Hwy 28 to the first of the Drum Lakes. A sign is posted on the south side of the highway. The trail begins right off the highway. For information contact Ruth Masters at 334-2270.

## Gold River

This community lies near the head of Muchalat Inlet on the west side of Vancouver Island and is near some of the finest limestone caves in Canada. For more information contact Karen Griffiths of Gold River (285-2691), an avid caver who has done much to publicize and promote this area.

### West Coast - Nootka Sound Area

The beaches of Hesquiat Peninsula from Escalante River to Estevan Point, and the west coast of Nootka Island, are of interest to hikers and naturalists. On the west side of Nootka Island a rough trail connects the Indian Reserve at Friendly Cove with Crawfish Beach about halfway up the Island. Sea caves and cliffs also occur on Nootka. The beaches are exposed sand and gravel backed with dense rain forest. In March and November gray whales migrate by. August brings the sea lions to the mouth of salmon streams, especially at the Escalante. Invertebrate sea animals abound in the rocky crevices.

**Plants**: Sitka spruce, hemlock, cedar, salal, evergreen huckleberry, red huckleberry and some yew.

Access is by boat via the government wharf 12 km west of Gold River. The coast is 30 km further west down Muchalat Inlet. A coastal freighter, *M.V. Uchuck*, departs from Gold River every Tuesday and Thursday for settlements in the Nootka Sound area. It carries passengers once a week to Friendly Cove. Contact the *Uchuck* at 283-2325.

### Friendly Cove (Yuquot) - Nootka Island

Vegetation consists of Sitka spruce/evergreen huckleberry forest. Arbutus trees grow along Muchalat Inlet, a western limit for this tree. Good salmon fishing may be had around the point.

**Birds**: bald eagle, raven, shorebirds, sea ducks, a few alcids, shearwaters and petrels.
**Mammals:** mink, river otter, marten, deer, bear and wolf.
**Marine Mammals: s**eal, sea lion, occasionally sea otter, humpback whales, finbacks, gray whales (in April) and blue whales (occasionaly).

Friendly Cove is reached by boat, either private or on the *Uchuck*, and by chartered float plane. The site is an Indian village and permission for visits or camping may be made by contacting the Mowachaht Band office at 283-2532. In case of emergency there is a radio phone in the lighthouse or ask Ray Williams to use his C.B. radio kept in one of the occupied native houses.

*Bob Sutherland and Ruth Masters*

## JOHNSTONE STRAIT (NORTH END) AND ADJACENT ISLANDS

The strait, lying on the northeast side of Vancouver Island, is the best place in the world to observe killer whales (orcas).

Killer whales are numerous, easily accessed, and fairly easy to find. The best time to see them is July to September. Pods of orcas cruise Johnstone Strait, Blackfish Sound and adjacent passages in their daily search for food (mainly salmon in summer). The orcas rub themselves along shallow pebble beaches on the south side of Johnstone Strait with activity centred on, but not restricted to, Robson Bight. Do not approach the whales too closely in boats or follow for prolonged periods of time — this is harassment. Orcas are used to boats and are not dangerous unless provoked. For centuries the area has been known for its abundance of whales. Prior to the commencement of commercial whaling, which ended with the closing of Coal Harbour Whaling Station in Quatsino Sound in 1969, fin whales and the occasional humpback and sperm whale came here. A few minke whales occur in the area of Blackfish Sound and Johnstone Strait.

Sea lions haul out on small islets in Queen Charlotte Sound west of Bonwick Island. Harbour seals haul out on small rocks in Knight Inlet south of Owl Island and Wedge Island.

A number of bird species can be seen here feeding on small fish chased to the surface by salmon (mainly pink and sockeye). This feed, sometimes called "herring balls," also attracts minke whales. Winter is the best season to spot coastal birds. To see both the whales and birds come in August.

Underwater sea life is abundant. Scuba and skin diving opportunities are excellent but hazardous in some locations because of the strong currents. Dive at slack tide at Stubbs Island, Blackney Passage, and Weybourne Passage.

The shore line consists of old growth forests in the lower reaches of Tsitka River. This is the only large old-growth forest on the east side of Vancouver Island.

**Birds:** bald eagle, storm petrel, sooty shearwater, rhinocerous auklet, common murre, Arctic loon, Bonaparte's gull, marbled murrelet, and an occasional Caspian tern. Many eagles are present.

**Further Information**

Access is by car ferry from Port McNeill to Alert Bay (Cormorant Island) and Sointula (Malcolm Island). Public boat ramps and docks are available at Alert Bay and Port McNeill. The closet ramp and docks to Johnstone Strait are privately maintained at Telegraph Cove which is accessible by road. The turn off is 10 km east of Port McNeill on Hwy 19. Follow the signs on the main logging access road past Beaver Cove to Telegraph Cove where there is a boat launch fee and daily parking fee. • Camping is not permitted inside Robson Bight, because the activity disturbs the orcas from using their rubbing beaches. Private campsite at Telegraph Cove Wilderness campsites are available in the strait. • Many whale watching charter services are available in Port McNeill or Telegraph Cove including Stubbs Island Charter - at 928-3185 aboard the "Lukwa". Guided sea kayak trips of the area are given in July and August by Ecosummers Expeditions, 1516 Duranleau St. Vancouver or phone them at 669-7741. Sailing charters may be had with Bluewater Adventures "Island Roamer" in Vancouver. • Further information may be obtained from Jim and Ann Borrowman at 928-3117 (see above). Nautical charts include 3596 and 3568. Those with small boats will need to carefully read current charts for Blackney Passage.

*Bob Sutherland*

# PORT HARDY

Situated at the north end of the Island Hwy, the area around Port Hardy provides a wide selection of natural history sites to explore.

A variety of birds can be found on Cape Scott Park's beaches and coves, where they stop on the way to and from Alaska and points south. Large mammals such as elk, deer, wolf and cougar have also been sighted in this park and the back country around Brooks Peninsula. Marine mammals are common on the coast and pods of gray whales and groups of sea lions are often seen passing through off the west coast. Resident killer whales find a home on the east coast of the island. Marine life is prolific along the rugged coast. It is well worth the time to search out tide pools for anemones, seastars and other creatures. Plant life is diverse, with alpine flowers on summits like Mt Cain and other high spots, and in woodlands, providing shows every spring and into summer.

## Cape Scott Provincial Park

This 15,070-ha provincial park, located at the very tip of Vancouver Island, is about a 48-km drive from Port Hardy to Holberg on the B.C. Forest Service Rd, and a fairly boggy seven-hour hike from the beginning of the trail to the beach. Follow the signs after reaching the Elephant Crossing, past the logging camp.

The hiking trail passes through thick forests, over Fisherman's River, past a lagoon and down to the sea and long stretches of sandy beach. Bird life is diverse, with the estuary on San Josef Bay being a prime feeding ground for a variety of birds. Offshore islands, including Triangle Island, contain the largest colonies of tufted puffins to be found in B.C. Sea lion rookeries are also found on offshore islands, including Scott Islands.

**Birds:** snipe, spotted sandpiper, albatross, rufous hummingbird, pileated woodpecker, Steller's jay, winter wren, and short-eared owl.
**Mammals:** deer, elk, wolf, cougar, river otter and black bear.
**Marine mammals:** gray whales, sea lions, seals, and the occasional harbour and elephant seal.
**Marine life:** clams, crabs, eelgrass, red rock crabs and nudibranchs.
**Plants:** red and yellow cedar, lodgepole pine, western hemlock, Sitka spruce and some true fir. Undergrowth: salal, salmonberry, huckleberry, ferns, mosses, and skunk cabbage.

An alternate trail from the end of the logging road through another part of the park may be taken to San Josef Bay. On top of Mount Saint Patrick at the far end of the beach in San Josef Bay is an interesting bog with shrunken trees and white gentians.

Holberg has a gas station and motel, but it is advisable to gas up at Port Hardy first. Check in at the Regional District Mt Waddington office in Port McNeill for further information.

## Georgie Lake

Canada Works crews built a 3.5 km trail from the campsite at this lake to Songhees Lake. The trail features bog vegetation, and an abandoned beaver dam. Purple gentians, shooting stars and butterwort line the shore of this lake; good cutthroat fishing in lake.

The Georgie Lake campsite is about 13 km from Port Hardy, off the Holberg B.C. Forest Service Rd.

## Hardy Bay and Beaver Harbour

These two local Port Hardy beaches offer a variety of sea life and seashore vegetation. The seawall along Hardy Bay provides an easy walk, but the more adventurous person will prefer to walk from the government wharf out toward the mouth of Tsulquate River.

Take a canoe out towards Masterman Islands where large red sea urchins line the rocky bottom on the way or explore the islands in Beaver Harbour and look for the historic kelp factory on Deer Island used by the Japanee before the second World War. Watch for winds from the southeast which cause very rough water for canoeists.

**Marine life**: anemones, crabs, seastars and moon snails.

For information on Hardy Bay contact the Port Hardy Museum at 949-8143.

## Marble River

The bluffs, rising from Alice Lake near here, offer a variety of violets and saxifrages. Numerous birds frequent the bushes and small islands around the lake. A loop trail begins at the far end of the upper campsite and goes down beside Bear Falls and beautiful gorges.

Canoeists can put in at the boat launch across from the campsite and paddle past the Marble River Chinook Hatchery into Alice Lake.

The campsite is only 15 km from the Port Alice junction off the Island Hwy and is considered by locals as one of the best sites to camp on the North Island.

## Little Hustan Caves

This cave park offers external rock arch formations and interesting cave features. Bedrock in the area is Quatsino Formation limestone and as a result unique forms of vegetation, that prefer sweeter or less acidic soil, grow here.

The site is a regional park adjacent to Anutz and Hustan Lakes. It lies off the Zeballos Rd, only 20 minutes off the Island Hwy south of Port McNeill.

## Mount Cain

This regional ski hill offers alpine hiking trails in the late spring and early summer. Look for dwarf juniper, white gentian, and mountain blueberry together with other alpine flowers. To reach the hill take the Schoen Lake cutoff from the Island Hwy, south of Woss and follow the signs on the relatively short drive along a logging road.

**Further Information**
Ministry of Forests officials have information on the area and maps may be picked up at their office — phone them at 956-4416. The regional district office in Port McNeill has information on most of these sites, call them at 956-3301. The Canadian Forest Products number is 281-2300 and Western Forest Products is 956-4446. • Special references on the area include: *The Cape Scott Story*, by Lester Peterson, 1974. Mitchell Press, Vancouver.

*Annemarie Koch*

# CHECLESET BAY KYUQUOT SOUND

This is the best spot in B.C. to find sea otters. There is a colony of tufted puffins on Solander Island and a colony of storm petrels on Thomas Island. Both islands are very exposed to the open Pacific.

The sea otters spend much of their life floating on their backs in salt water kelp beds and are most common around Bunsby Islands and Acous Peninsula. Harbour seals and sea lions haul out on O'Leary Islets. Black-tailed deer are especially visible on Bunsby Island. Black bear are seen on Vancouver Island proper.

**Further Information**
Access is by logging roads and by boat. Use caution on these roads and watch for logging trucks. To reach Fair Harbour, drive north on Hwy 19 to Woss. The turnoff to Zeballos is about 25 km west of Woss Camp. Look for the road sign to Zeballos. Drive about 35 km west of Zeballos to reach Fair Harbour where there is a boat launch. There is a government wharf and floats at Kyuquot, but it is not accessible by road. An alternative is to charter a float plane out of Gold River or Port McNeill. Or you may come by sea on the weekly sailing of the *Uchuk II* cargo ship out of Gold River. • The hotel is open at Zeballos, and there is bed and breakfast in Kyuquot. There are no formal campsites on the road in. Boaters using a water route will find numerous spots in the area, with lots of fresh water except on Bunsby Islands. No facilities are available on the open coast for 65 km north of Kyuquot or until you reach Winter Harbour. On the way in, when leaving Hwy 19, fill your gas tank at Woss Camp. Carry a spare tank of gas or a locking gas tank. Inquire locally about a safe place to leave your vehicle. • Nautical charts include 3682 and 3683.

*Bob Sutherland*

# SOUTHWEST BRITISH COLUMBIA

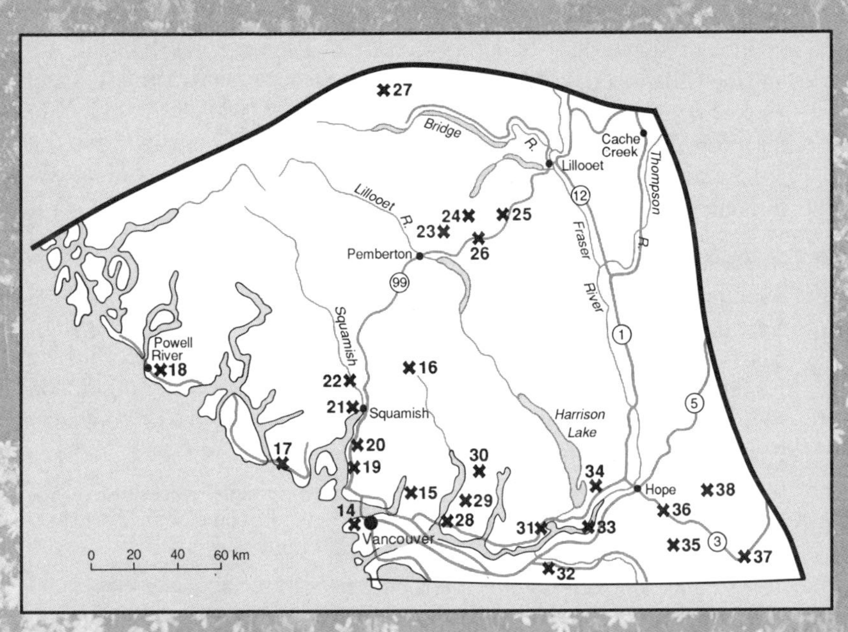


14. Vancouver Area
15. Mount Seymour P.P.
16. Garibaldi P.P.
17. Sunshine Coast
18. Powell River
19. Horseshoe Bay to Pemberton: Hwy 99
20. Porteau Cove P.P.
21. Squamish
22. Alice Lake P.P.
23. Birkenhead Lake P.P.
24. Pemberton Valley
25. Joffre Lakes/Peak Area
26. Pemberton to Lillooet via Duffey Lake Road
27. Southern Chilcotin Mountains
28. Pine Mountain Ecological Reserve
29. University of British Columbia Research Forest
30. Golden Ears P.P.
31. Harrison Mills/Nicomen Slough
32. Cultus Lake P.P.
33. Agassiz - Maris Slough
34. Sasquatch P.P.
35. Skagit Valley
36. Hope to Princeton - Hwy 3
37. Manning P.P.
38. Cascade Wilderness

# VANCOUVER AREA

The city of Vancouver and surrounding communities have retained large blocks of green space ranging from a major zoo to a point of land with huge trees that have never been logged. These regional parks are havens for wildlife and wild flowers.

## Crippen Regional Park on Bowen Island

Set on an island just west of Horseshoe Bay ferry terminal, this park offers tidal flats, coastal rain forest, arbutus groves, and wetlands. Special features include a fish ladder on Killarney Creek, fish hatchery, marsh boardwalk and the Killarney Lake Nature Trail. To reach the park take the ferry from Horse Shoe Bay to Snug Cove on Bowen Island and walk the one block to the information kiosk at the historic Union Steamship Company store. The Killarney Lake Loop Trail, one of at least three trails in the park, passes through a flooded forest and marsh rich in bird life.

## Lighthouse Park

On a point of land along the southwest side of the north shore of West Vancouver, the park extends south into the bay. The 75 ha of land preserve a last remnant of primeval forest that once covered the coastline. Bird life is abundant. The granite rocks throughout the park are over 100 million years old. Giant Douglas-fir around 400 years old, up to 61 m high and 2 m in diametre dominate the smaller trees.

**Plants:** grand fir, western yew, red alder, broadleaf maple, western hemlock, Douglas-fir, Western red-cedar, western sword fern (most common), spiny wood and lady ferns.

To reach the site, drive across the Lions Gate Bridge and take the second exit to travel west on Marine Drive for about nine km. Watch for signs on the left (south), to the park. Phone the park at 922-5408, or 922-8014 for further help or to find where to obtain the book about this site. An excellent reference recently published on Lighthouse Park is *Nature West Coast* by Kathleen M. Smith, Nancy J. Anderson, and Katherine I. Beamish, 1988, Sono Nis Press, Victoria.

## Capilano River Regional Park and Fish Hatchery

The park contains magnificent Douglas-firs, 300 years old, with a long canyon cutting through granite. The 35-km-long Capilano River begins in snow at 1,524 metres high in the mountain and plunges over Cleveland Dam into the canyon. Special features include the swift-flowing river, coastal rain forest, waterfalls, granite cliffs, fish hatchery and good interpretive exhibits. Across the bridge from the hatchery is a remaining giant fir. To reach the park from the upper level of Hwy 1, take the Capilano Rd/Grouse Mountain exit and drive north to the fish hatchery turnoff or continue to the Cleveland Dam parking lot.

## Lynn Headwaters Regional Park

This site offers rugged wilderness with an uncontrolled, often swift-flowing creek, coastal rain forest, mountain forest and subalpine meadows, all just at the edge of North Vancouver. Hiking trails, some of which have interpretive pamphlets, are shown on a brochure available at the entrance.

To reach the park from the Second Narrows Bridge, take Hwy 1 to Lynn Valley Rd in North Vancouver and follow this road to the park entrance. Lynn Canyon Ecology Centre in Lynn Canyon Municipal Park features a Visitor Centre which is open daily. The park has a suspension bridge affording good views of the vegetation in the canyon.

## Stanley Park

This park is one of the best birding areas for winter waterfowl around the lower mainland(60-70 species). The greatest variety occurs from mid-November to mid-March with rarities showing up continually. Operated by the City of Vancouver, the 2470 ha park contains a zoo with a good collection of exotic birds and mammals, the Public Aquarium, extensive cultivated gardens, a large lake and large blocks of natural woodland. The best way to explore is to walk the numerous trails which criss-cross through the forest. Beneath the Lions Gate Bridge look for the recently established colony of pelagic cormorants, glaucous-winged gulls and pigeon guillemots. Good views can be had from the seawall walk along the edge of the park. The best time is June and July when eggs and chicks are present.

## Pacific Spirit Regional Park (University Endowment Lands)

This is Vancouver's most recent park (dedicated in 1989) with 765 hectares of forest and foreshore. Forests combine good examples of early successional alder and aspen stands with second growth conifers, together with old-growth Douglas-fir forest. At Camosan Bog, a sphagnum bog, look for sundew, cloudberry and lodgepole pine with sphagnum plants that may be 10,000 years old. Trails and other highlights are shown on the park brochure available through the Regional District office at 432-6350. A recent, helpful reference is *An Illustrated Flora of the University Endowment Lands* by G.B. Staley and R.P. Harrison, 1987, U.B.C. Press. The park is located in Point Grey adjacent to the University of British Columbia. Take Marine Drive, Chancellor Boulevard or 16th Ave to reach various park entrances.

## Richmond Nature Park

This site has a Nature House open all year with exhibits, blueberry bog, waterfowl viewing ponds with a board walk, and nature trails. Several nature programs are offered together with field trips to local natural areas. For program information phone 273-7015. The park is located at 11851 on the Westminster Hwy in Richmond.

## Burnaby Lake Regional Park

This nature park, comprising 300 ha of lake, marsh and lowland forest, sits in the centre of Burnaby municipality. Although wholly managed, it retains an essentially wild character.

**Birds:** black-headed grosbeak, rufous-sided towhee, red-tailed hawk, rough-winged swallow and four other species of swallows, black and Vaux's swifts, common nighthawk, wood ducks, osprey, Virginia rail, sora, pied-billed grebe, and double-crested cormorant.
**Mammals**: beaver and raccoon.
**Plants:** iris, bulrush, cattail, purple loosestrife, rushes, grasses, sedges, mannagrass, woody shrubs, Sitka willow, Pacific willow, hardhack, red alder, paper birch, western hemlock, false lily-of-the-valley (May to early July), red-cedars, Alder (common), cottonwood, Western red-cedar and big-leaf maple.

The park is located 20 km east of the centre of the city of Vancouver between the Trans-Canada and the Lougheed highways. Leave the Trans-Canada Hwy either on Kensington, the road on the west park boundary, or Cariboo Rd on the east boundary of the park. The Burnaby Lake Nature House on the north shore of the lake offers programs and information from May to September. They have maps of the park and trails. Phone: 432-6350.

## Belcarra Regional Park

This park offers 9 km of mud, sand and rocky shoreline and one of the region's warmest lakes, Sasamat Lake. Coastal rain forest offers woodland birds. Located in North Vancouver, the park is reached by taking the turnoff from Barnett Hwy onto Ioco Rd, just east of Port Moody, and following the signs through Ioco to White Pine Beach or the Belcarra Picnic Area.

## Minnekhada Regional Park

The park has marsh, cedar forests and coastal rain forests. Features include a man-made marsh rich in waterbirds and mammals. The area has numerous hawks and song birds. Hiking trails wind through cedar and fir forests onto rocky knolls. Take the Lougheed Hwy, turn off on Coast Meridian Rd, and follow the directional signs.

## Campbell Valley Regional Park

Campbell Valley Park lies east of Vancouver and south of Langley. The filbert (hazel) nut orchard is the best birding site in the park. The vegetation ranges from formerly cultivated open fields to forests.

**Birds**: brown creeper, purple finch, red-breasted nuthatch, orange-crowned warbler, black-capped chickadee, northern oriole, bushtit, American goldfinch, cedar waxwing, kinglet, downy woodpecker, savannah and white-crowned sparrows, swallows, red-winged blackbird, snipe, waterfowl, yellow warbler, marsh wren, sora, Virginia rail, common yellowthroat, great horned, barred, barn, western screech, saw-whet, and pygmy and long-eared owls.
**Amphibians**: red-legged frog, Pacific tree frog, red-backed salamander, rough skinned newt and boreal (western) toad.
**Butterflies**: Milbert's tortoise shell, western swallowtail and Lorquin's admiral.
**Wildflowers**: tall fringe cup, piggy-back plant, western bleeding heart, yellow violet, trillium and broad-leaved aven.

To reach the park from the Trans-Canada, take the Langley 200th St south. Drive 14.5 km and turn east on 16th Ave for the North Valley Entrance. Alternatively, from Hwy 99 South, take the 8th Ave East exit and drive 7.4 km to the South Valley Entrance. The Visitors' Centre is open weekends, May-September. Phone: 432-6350.

*Jude and Al Grass*

## Wetlands of the Lower Fraser Valley

The Fraser estuary/delta supports more wintering waterbirds, shorebirds and raptors than any other place in Canada. One million waterfowl and over a million shorebirds pass through each fall and spring. At the peak of migration, in November, 200,000 dabbling ducks are present.

The diving ducks throng to the shores of Boundary Bay and around Stanley Park.

About 38 species of fish have been found downstream from Hope. The five species of salmon are the most prized.

Almost every day of the year fish migrate through the estuary. At peak times, between early March and late May, they may number a million passing per day.

One of the largest salmon runs in the world occurs on the Fraser River. The chinook spawning runs begin in late spring and peak in early autumn. Spawning runs of the Coho begin in late summer and peak in September. The pink runs every two years with about six million fish coming up between mid-August and mid-October.

The annual run of sockeye is one of the world's largest with over ten million fish coming up the river between June and September.

Chum enter the main arm of the river after the third week of September and stay in the estuary for one to four weeks. There are usually two major peaks in the run, beginning in mid-October and in early November.

Spawning runs of sea-run cutthroat trout occur between November and mid-April. Steelhead trout run mainly between November and May.

Tidal marsh plants include cattail, sedge, bullrush, saltwort, saltgrass and arrowgrass.

Peat bogs are the only significant remaining examples of native vegetation within the dykes. These are thick mats of sphagnum in various stages of decomposition. They keep the water table high through capillary action and produce organic acids that help to maintain a high water acidity and discourage decomposition. Organic material accumulates faster than it can rot which in turn results in the bogs reaching rising higher than the surrounding area.

**Plants**: Labrador tea, salal, blueberry, bog cranberry, lodgepole pine, birch, hardhack, Western red-cedar, western hemlock and Sitka spruce.
**Birds**: pintail, wigeon, mallard, green-winged teal, wood duck, blue-winged, cinnamon teal, three scoters, mergansers, both goldeneyes, scaup, harlequin, bufflehead, ruddy duck, red head and canvasback, tundra and trumpeter swans, feral Canada goose, Taverner's Canada goose, lesser snow goose, brant, great blue heron, bitterns, green-backed heron, black-crowned night heron, sandhill crane (rare), bald eagle, red-tailed hawk, northern harrier, rough-legged hawk, gyrfalcon, peregrine falcon, merlins, kestrels, sharp-shinned hawk, Cooper's hawk, and short-eared, barn and snowy owls.
**Marine Mammals**: killer whales, harbour seals, sea lions and seals.
**Mammals**: muskrat, beaver, squirrels, flying squirrels, beaver, mink, river otters, red fox, raccoons, skunks, opossum, coyote, shrew moles, Townsend's voles, Townsend's chipmunks, mountain beaver, blacktail deer and black bear.
**Fish**: Cutthroat and steelhead trout, herring, Dolly Varden, mountain whitefish, white sturgeon, and eulachon.
**Marine life:** shrimp, ghost shrimp; butter, horse, little-necked and mud clams; cockles, mussels and three types of oysters.

## Sturgeon and Roberts Banks and Reifel Bird Sanctuary

The foreshore marshes and intertidal flats on these banks cover over 13,000 hectares and extend several kilometres into Georgia Strait. Close to the dykes they support thick stands of cat-tail, sedges, and bulrush that attract large winter flocks of lesser snow geese and other waterfowl. The open flats provide feeding habitat for large gatherings of shorebirds. Shorebirds, including rarities, can be viewed at close range around the sewage lagoons on Iona Island, a 'mecca' for Vancouver birders. The largest number of rare birds seen in B.C. occur here. Most of Westham Island is private property. But the northern portion, Reifel Island, is owned by the Canadian Wildlife Service. Here the George C. Reifel Migratory Bird Sanctuary and the adjacent Alaksen National Wildlife Area have helped to maintain Reifel Island as the richest area for wildlife in the delta. Over 230 species of birds have been recorded at the Reifel Sanctuary. About 20, 000 lesser snow geese that nest on Wrangel Island in the Soviet Union, winter on foreshore of Robert Banks facing the Sanctuary. The best time to visit the Sanctuary is between October and March when migrants abound.

**Birds**: mallard, blue-winged, cinnamon and green-winged teal, ruddy duck, shoveler, pintail, great blue heron, green-backed heron, American bittern, black-crowned night heron and numerous raptors, gulls, cormorants, grebes, shorebirds, and waterfowl, including swans.

To reach the north end of Roberts Bank and the Reifel Sanctuary cross the Fraser's south arm on Hwy 99 through the Massey Tunnel and turn west through Ladner to Westham Island.

## Burns Bog

This area contains 10,000 hectares of sphagnum peat, of which 2,000 hectares have been dug out for commercial purposes, making lakes. Bog plants thrive here and black bears and sandhill cranes are found in small numbers. The garbage dump on the southwestern corner of the bog attracts huge flocks of gulls and other scavengers including bald eagles and ravens. To reach the bog turn north from the Ladner Trunk Rd on 104th Street, and then west on the frontage road parallel to the north side of Hwy 99.

### Boundary Bay

This large bay was formed by tidal flows in the post-glacial period when the lower Fraser valley was a fiord. Its waters are much more saline than those found on the Sturgeon and Roberts Banks. So the plant life is halophytic, including eelgrass, sea lettuce, various algae, salicornia and several sedges and beach grasses. This abundance of vegetation attracts thousands of brant and other waterfowl. Boundary Bay dyke is about 16 km long and provides an excellent platform from which to scan the bay for the wide variety of water birds and birds of prey that inhabit the foreshore and tidal waters. To reach the site follow Hwy 10 east of Ladner and turn south on any of the streets between 64th and 112 Street that lead to the dyke on Boundary Bay.

### Mud Bay

Lying east of Boundary Bay, Mud Bay contains deltaic marshes between the estuaries of the Serpentine and Nicomekl Rivers. The bay and nearby shoreline are especially valuable as feeding and loafing habitat for waterfowl, shorebirds, raptors and harbour seals. The bay lies east of the intersection of Hwys 10 and 99.

### Serpentine Fen Wildlife Management Area

The farm lands, extending from Mud Bay to Langley, provide important winter feeding sites for thousands of ducks, shorebirds and birds of prey. Dyke walks and observation towers give good views over the meadows and marshes.

**Birds**: Feral Canada goose, lesser snow goose, swans, Wilson's phalarope, avocet, yellow-headed blackbird, short-eared and barn owls.
**Mammals**: muskrat, skunk, opossum, raccoon and coyote.

The area is reached via Hwy 99 or King George Hwy; turn on 44 Ave at the nursery and drive 200 m to the parking area on the left.

### Blackie Spit, Crescent Beach

This municipal park, on the south side of the Nicomekl estuary is a good place to spot migrating shorebirds, ducks, thousands of overwintering dunlins and gatherings of over 100 great blue herons in Mud Bay.

**Birds**: brant, dunlins.

To reach the spit, take Hwy 99 to White Rock-Crescent Beach turn off, and drive along Crescent Rd to the north side of Crescent Beach.

### Crescent Beach - Ocean Park - White Rock

This is a good area to see harlequin ducks, bald eagles, and other seabirds.

**Marine mammals**: harbour seals, gray and killer whales.

**Marine life:** seastars, hermit crabs, sand dollars and several interesting snails.
**Mammals:** northern flying squirrels and shrew moles.

This area is most easily reached by way of the "1001 Steps" at Kwomais Point. The steps are located at the foot of 15A Ave off 126A Street, Ocean Park. The White Rock pier on Marie Drive also provides a good vantage point for viewing sea birds on Semiahmoo Bay.

### Campbell River

The estuary, south of White Rock, attracts gulls, terns and black turnstones. Its upper reaches are the site of Campbell Valley Regional Park. Watch for small populations of beaver, wood duck, green-backed heron, Virginia rail and sora. Drive east from White Rock on 8th or 16th Ave to reach it.

### Surrey Bend

This flat apron, containing swamp, bog, forest and marsh, is all that remains of one of the largest undyked areas of the Fraser floodplain. When the river rises in flood, the Bend provides a "safety valve" or floodway. The varied plant life and rich population of birds and mammals make it a living museum of the Fraser Valley before white settlement. To reach it drive to the north end of 176th Street, or the east end of 104th Ave, in Surrey.

The ferry at the east end of 104th Ave takes visitors to Barnston Island where the dyke road offers a pleasant circuit of farmland and river shore habitats.

### The Pitt Valley

The whole flood plain of the Pitt Valley was formerly peat bog and marsh that provided wapato, or Indian potato, to the Katzie Indians and to geese and swans. Although most of the area has been drained, over 1200 ha of marsh and bog remain in the Pitt Wildlife Management Area. The site preserves a place for coast black-tailed deer, black bear, cougar, bald and golden eagles, osprey and sandhill crane.

On the west bank of the Pitt River, the Addington Marsh Sanctuary and Minnekhada Regional Park in Port Coquitlam offer ready access (via Prairie and Devon Rds) to a variety of marsh and forest species of wildlife.

To reach the valley drive east from Vancouver on Hwy 7; cross the Pitt River; fork left on Dewdney Trunk Rd; turn left (north) on Harris; and then east on McNeil Rd.

**Further Information**

A variety of publications are available from the Vancouver Museum and Planetarium Gift Shop at 1100 Chestnut Street, Vancouver (604) 731-4158 including bird lists, *Explore the Fraser Estuary* by Peggy Ward, Lands Directorate, Environment Canada, 1980, and *Waterfowl on a Pacific Estuary* by Barry Leach, B.C. Provincial Museum, 1982. The White Rock and Surrey Naturalist Society has a checklist of about 165 birds, which is available from them at P.O. Box 44, White Rock, B.C. V4B 4Z7.

*Barry Leach and Jack Williams*

# MOUNT SEYMOUR PROVINCIAL PARK

Within an hour's drive of downtown Vancouver, Mt Seymour Provincial Park offers wet forests that grade to subalpine meadows with 117 bird species including blue grouse, black and Vaux's swifts, red-breasted sapsucker, gray and Steller's jays, black-throated gray and MacGillivray's warblers, rosy finch and rock ptarmigan. The forest is typical of the coastal forest found on the west side of the province.

**Birds**: Hutton's Vireo, black-throated gray and Townsend's warblers, rufous hummingbird, western tanager, Swainson's thrush, winter wren, red-breasted sapsucker, pileated woodpecker, red-tailed hawk, Cooper's hawk, goshawk, willow flycatcher, MacGillivray's and Wilson's warblers, black and Vaux's swifts, gray jay, raven, varied thrush, blue grouse, pygmy owl, water pipit, and hairy woodpecker.
**Plants:** Salmonberry, vine and broadleaf maple, cascara, western hemlock, red huckleberry, trailing rubus, white birch, black cottonwood, Western red-cedar, western yew, salal, bunchberry, deer fern, amabilis fir, yellow cedar and beak moss.

The Alpine Trail to Mystery Peak, Pump and Mt. Seymour is a strenuous hike and at least an 8-hour return trip. In the fall, hawks migrate near the peak.

Goldie Lake self guiding nature trail takes off from the east side of the parking lot. It is an easy 2 km hike through subalpine forest. Brown salamanders are found here.

**Birds**: spotted, solitary sandpipers, chestnut-backed and mountain chickadees, golden-crowned kinglet, red crossbill, purple finch, blue grouse, and hermit and varied thrushes.
**Plants**: copperbush, false azalea, blueberries and huckleberries, arctic star-flower, grass-of-parnassis, leptarhenra and false asphodel.

Somewhere between 100 and 140 million years ago volcanoes rose from the sea to form these mountains. Near the lower chairlift these ancient lavas are exposed including angular fragments that were caught in the still liquid lava. Near Brockton Point, a rocky outcrop on Mt Seymour, there is a band of volcanic ash weathering whitish in colour. The ash was laid down in fine-grained beds and is believed to have been deposited in quiet water.

**Further Information**
From downtown Vancouver drive 15 km northeast, via the Second Narrows Bridge, to Mt Seymour and on to the Park. • The park brochure with a map showing trails and ski areas is available from Provincial Parks, 1610 Mt. Seymour Rd, North Vancouver, B.C. V7G 1L3.

*Al Grass and Vic Adamo*

# GARIBALDI PROVINCIAL PARK

A wilderness area awaits you within a two hour drive from downtown Vancouver. Come to see spectacular scenery including glaciers, volcanic features, sub-alpine and alpine meadows. Many birds are mainly restricted to high elevations or high latitude.

**Birds**: white-tailed ptarmigan, three-toed woodpecker, horned lark, Clark's nutcracker (common), gray jay, mountain and boreal chickadee, rosy finch, red and white-winged crossbill and pine grosbeak.

## Black Tusk Area

This area, on the north side of Garibaldi Lake, has spectacular scenery. A checklist shows 125 species of birds, including those found along the 9 km Garibaldi Lake Trail.

**Birds**: yellow-rumped and Wilson's warblers and white-crowned sparrow, Clark's nutcracker, gray jay, chestnut-backed chickadee, rufous hummingbird, three-toed woodpecker, Osprey, American dipper, spotted sandpiper, white-tailed ptarmigan, blue grouse, mountain chickadee, rosy finch, red and white-winged crossbills, golden-crowned sparrow, water pipits, pectoral and Baird's sandpipers, greater and lesser yellowlegs, golden eagle and northern harrier.
**Mammals**: pika, hoary marmot, Douglas squirrel, northwestern chipmunk, white-footed deer mouse, snowshoe hare, pine marten, short-tailed weasel, mountain goat, little brown bat, black-tailed deer.
**Fish**: rainbow trout.
**Trees**: sub-alpine fir, mountain hemlock, whitebark pine and yellow cedar.
**Wildflowers**: avalanche lily, spring beauty, western anemone, sitka valerian, arctic lupine, Indian paintbrush, fan-leaf cinquefoil, cow parsnip, mountain daisy, tall brook ragwort, pink mimulus, white rein orchid, fringed grass-of-parnassus, common fireweed, alpine fireweed, silky phacelia, Indian paintbrush, white paintbrush, common butterwort, white rein orchid, cascade stone crop, and pearly everlasting.

## Diamond Head Area

The area, which includes Mt. Garibaldi, Diamond Head, the Opal Cone and Garibaldi Neve, is noted for extensive expanses of red and white heather. Lands around the Opal Cone are rich in sub-alpine wildflowers. Birds and mammals are similar to those found in the Black Tusk area.

## Cheakamus Lake Area

This is a beautiful glacier-fed lake, at about 830 m, framed by high peaks and ridges.

The access trail to the lake passes through a forest of giant western hemlock, Western red-cedar and Douglas-fir.

**Birds**: Barrow's goldeneye, common merganser and common loon, barred owl, rufous hummingbird, Vaux's swift, northern flicker and varied thrush.
**Fish**: Rainbow trout, Dolly Varden.

## Singing Pass Area

The trail to the pass is through forest at lower elevations and through sub-alpine flowering meadows at higher elevations. Birds include northern pintail, American wigeon and American dipper on Russet Lake. Mountain goats may be seen regularly.

### Further Information

Hwy 99 will take you north from Horseshoe Bay. Side roads link this highway to parking lots from which you can hike into Diamond Head, Black Tusk, Cheakamus Lake and Singing Pass. • Diamond Head Area may be reached from the access road 49 km north of Horseshoe Bay (4 km north of Squamish). Drive in 16 km from Hwy 99 to the parking area and trail head. The hiking trail to Elfin Lakes is 11.2 km with an elevation change of 600 m. • The Black Tusk Area access road lies 78 km north of Horseshoe Bay (37 km north of Squamish). Take the 2.4 km paved road from Hwy 99 to the parking lot and trail head. You may also take the train on the British Columbia Railway to Garibaldi Station which is 4 km southwest of the parking lot. The trail to Garibaldi Lake is 9 km with an elevation change of 920 m. • To reach Cheakamus Lake, drive north 97 km from Horseshoe Bay (48 km north of Squamish). An 8 km logging road joins Hwy 99 with the parking lot. The hiking trail to the lake is 3.2 km with minimal elevation change. • The Singing Pass Area access road lies 104 km north of Horseshoe Bay at Whistler Village. Take the logging road that leads from behind the village to the trail head about 4.8 km to the south east. The hiking trail is 6.4 km to Singing Pass and a further 2 km to Russet Lake. • Campgrounds for tenting are located at Diamond Head and Black Tusk Areas. A shelter is located at Russet Lake. The Elfin Lakes shelter in the Diamond Head Area is available for accommodation year-round. • A reference book on the park is: Roberge, Claude. 1982. *Hiking Garibaldi Park at Whistler's Back Door.* Douglas and McIntyre Ltd., Vancouver. • Garibaldi Provincial Park map is available all year at the District Office in Alice Lake Park, and in summer at Garibaldi Lake and some tourist offices. • Panorama Ridge Interpretive Trail brochure and a Checklist for the Birds of Black Tusk Area are available, in summer, at Garibaldi Lake.

*Barb McGrenere*

# SUNSHINE COAST

Located in the rain shadow of Vancouver Island, the Sunshine Coast is well named. The area is famous for spectacular scenery with mountains, fjords, islands, beaches and forests. Other highlights include the Skookumchuck, a tidal rapids best viewed at low tide. Nearby Princess Louisa Inlet is considered by many to be a world renowned beauty spot. There is good diving in the area and excellent salmon fishing is available.

The checklist of birds for the area lists 241 species, with 17 being casual or accidental. Marine and shorebirds are the local specialties, and they are best spotted in winter. Notable water birds include parasitic jaegers present in mid-September to mid-October, and Rosy finches are found above the 1700 metre level.

The low elevation forests on this coast are a mix of conifers and deciduous trees, and white fawn lilies provide beautiful displays on mossy rock outcrops in April.

Painted turtles may be observed sunning themselves on submerged logs at Paq Lake, beside the highway in Pender Harbour.

**Birds**: loons, cormorants, grebes, herons, ducks of all kinds, a variety of shorebirds, gulls, alcids and bald eagles. Thirteen species of gulls and terns have been recorded with glaucous-winged gull the most common, eight warbler species, Hutton's vireo, Black-throated gray warblers, Black and Vaux's swifts, black oystercatcher, black turnstone, surfbird, and rock sandpiper

**Marine mammals**: harbor seals, Northern or Steller's sea lions, porpoises and killer whales.

### Mission Point on Davis Bay

This is the best spot to find shorebirds mentioned above. The pebbly spit of Mission Point is probably the most reliable spot in B.C. to see these species and the more uncommon gulls. Look for alcids out on the water. Marbled murrelet are abundant in winter and frequent in summer. Common murre and pigeon guillemot are frequently seen too. Davis Bay lies on the first stretch of highway fronting on the ocean, 22 km from the Langdale ferry terminal. Park your vehicle where the highway first meets the ocean and walk back south for about 100 metres to the mouth of Chapman Creek.

### Wilson Creek Estuary

This site is excellent for a wide variety of shorebird species from early July to the end of September. Walk around the west side of the mudflats to reach the ocean. Many rarities have been found here. To locate the site, head about a km back from Davis Bay towards Gibsons, park near the Chevron service station. Walk in at low tide to find the mud flats where migrant shorebirds congregate.

### Porpoise Bay

Come here in winter and spring for a variety of waterbirds. On the left lies the Sechelt Marsh which is good for waterbirds in winter and passerines in summer. The bay is reached by turning north off Hwy 101 at the only traffic signal in Sechelt and proceed one km to Porpoise Bay.

**Further Information**

To reach this coast take the 35 minute ferry ride from Horseshoe Bay, just northwest of Vancouver, to Langdale, and drive north on Hwy 101. • *Sunshine and Salt Air - A Recreation Guide to the Sunshine Coast*, describes many hikes and trails on this coast.

*Tony Greenfield*

# POWELL RIVER

Isolated from the rest of the coast by fjords, the area lies off the beaten track, but is only 150 km by road and ferry from Vancouver. Cranberry Lake, a natural area within the community, attracts water birds. MacMillan Bloedel Ltd. have an extensive system of over 160 km of logging roads which are open for weekend use, with prearranged permission.

All the typical west coast birds are present with occasional migrants from Asia showing up. Gull and cormorant colonies occur off shore.

**Marine mammals**: sea lions, seals.
**Mammals**: coast black-tailed deer, otter, mink, and Douglas' squirrel.

Two nature trails have been developed by the town. To reach the top of Mount Valentine View Park drive to the corner of Cranberry and Crown and turn up the hill on Crown to the end of the paved road. Park the car and walk up the trail, up a rock staircase to the view. A sign then directs the hiker on a 600 metre trail across the mountain top with stairs and tables along the way.

The other trail, "Willingdon Creek Nature Trail," begins opposite the Willingdon Beach campsite and runs for about 1.5km to Joyce Ave. It is an easy hike of an hour with winding staircases, a bridge, and "The Hobbit," a hole in the ground.

**Further Information**

Two major logging roads lead through forests, up mountains to beautiful lakes. These roads start at Lang Bay on the west side and Stillwater on the east side of the river flowing out of Lois Lake. Maps and a recreational guide can be picked up at the office of the Stillwater Division of Macmillan Bloedel office at 4727 Marine Drive near Eaton's store. • The community is virtually an island with access on one road from the south, Hwy 101, that requires two ferry crossings, or by boat, sea and air. Take the Longdale Ferry from Horseshoe Bay and follow Hwy 101 along the Gold Coast. Alternatively, cross the Strait of Georgia on the ferry from Little River near Courtenay. Maps of the area showing roads, contours and communities include 92F-15; 92F-16E & W; 92K-2.

# HORSESHOE BAY TO PEMBERTON: HWY 99

This all weather road passes from forests to the high country and gives access to sub-alpine and alpine areas close to Vancouver. Also it is the road to the Whistler ski and winter playground area.

**Km 0.0:** At Horseshoe Bay turn off on Hwy 99.

**Km 0.4:** Roadside pulloff on the left. Look for common raven and Steller's jay. In winter at the marina and ferry terminal surf scoter, Barrow's goldeneye and bufflehead are seen.
**Plants:** broadleaf maple, Douglas-fir, arbutus, mountain ash and stonecrop.

**Km 24.7:** Entrance on the left to Porteau Cove Provincial Park.

**Km 36.0:** Entrance to Murrin Provincial Park on the left. This park has a small lake has yellow pond lily, buckbean, and skunk cabbage at the shallow southwest end.
**Plants:** Western red-cedar, western hemlock, salal, salmonberry and thimbleberry.

**Km 40.7:** Shannon Falls Provincial Park on the right (east) side of the road. The waters of Shannon Creek plummet down 120 m cliffs lined with Western red-cedars and Douglas-firs. The falls are spectacular after a heavy rain and during the alpine snow-melt period. Vine and broadleaf maples can be found below the falls. Salmonberry also grow nearby.

**Km 44.7:** Traffic lights; continue straight through to Pemberton.

**Km 47.8:** Bridge across the Mamquam River. During the winter bald eagles can be seen in the trees lining the river. American dippers feed in the shallow waters along the riverbank and common mergansers can be observed on the river.

**Km 48.6:** Turn off on the right to the Diamond Head area of Garibaldi Provincial Park.

**Km 59.3:** Brohm Lake parking lot on the left. The south end of the lake has cattails and pond lilies. Shore pine, Douglas-fir, western hemlock and Western red-cedar are the dominant trees along the lakeshore.

**Birds:** chestnut-backed and black-capped chickadees and warblers in migration.

**Plants:** skunk cabbage, salal and licorice fern.

**Km 75.8:** Stop of interest on the right. The sign warns of the hazard caused by unstable volcanic rock of the Barrier, a large rock wall in Garibaldi Provincial Park.

**Km 78.5:** Black Tusk/Garibaldi Lake area turnoff.

**Km 86.8:** Entrance to Brandywine Falls Provincial Park. A short walk from the parking lot takes you to the 70 m waterfall. The waters of Brandywine Creek pour over a great basalt sill and fall 66 m into a deep pool at the bottom of the gorge. American dipper can be found in the stream above the falls. Pikas live in the broken rock slopes along the falls trail and along the railway which crosses the park. Look for Douglas squirrels too. Maidenhair ferns and mosses cling to the chasm walls. Understorey plants along the falls trail include bunchberry, fireweed, bracken fern, false box, and red huckleberry.

**Km 91.7:** Columnar basalt outcrops along the right side of the road. These columns were formed by the rapid cooling of molten lava under the water. Pearly everlasting, fireweed, falsebox and penstemon can be found growing in this rocky area. Small cottonwood saplings grow here.

**Km 110.5:** Pull off on the right allows you to spot Penstemon blooming on the bluffs above the road in spring. Look for migrating waterfowl on Green Lake in spring and fall.

**Km 134.3:** Nairn Falls Provincial Park entrance with a relatively dry forest as compared to forests around Howe Sound. Douglas-fir, Western red-cedar, western hemlock and Pacific dogwood are the main trees. Undergrowth consists of a mixture of Oregon grape, rose, thimbleberry, false box, star flower, pipsissewa and old man's beard. The round trip hiking distance to the falls is about 5 km. The waters of the Green River come crashing over a double set of falls in a narrow gorge, making a right angle turn to get around a rock barrier between the upper and lower falls.

**Km 136.3:** One Mile Lake and picnic area turn off. A trail has been developed part way around the lake. The Eastern kingbird is here in summer. During migration watch for pied-billed grebe, American wigeon, wood duck and several warbler species. The forest consists of Douglas-fir, Western red-cedar and some mountain ash.
**Plants**: (understory) false box, thimbleberry, snowberry, oak and bracken ferns, Oregon grape, bunchberry, pipsissewa, large wintergreen, rattlesnake plantain, queen's cup, twinflower and fireweed; (lake) yellow pond lily and cattail.

*Barb and Mike McGrenere*

# PORTEAU COVE PROVINCIAL PARK

The park lies on the shores of Howe Sound, a coastal fjord formed by the action of glacier ice. The site provides an opportunity to explore the foreshore of Howe Sound in an area where there is little public access to the ocean. The moderate depths and minimal currents at Porteau beach have made it popular with divers. To provide a solid substrate for sea life, two ships and some concrete rubble were sunk in the cove in 1980 and another in 1985.

The park lies in mixed coniferous-deciduous forest.

**Marine Life**: (intertidal zone) blue mussels, California mussels, common acorn barnacles, sitka and checkered periwinkles, shield limpets, purple shore crabs, green shore crabs, red rock crabs, hairy hermit crabs, black pricklebacks, rock pricklebacks, tidepool sculpins, blenny species, gunnel species, mottled stars and purple sea urchins; (subtidal) orange sea pens, plumose anemones, swimming sea anemones, California sea cucumbers, armoured sea cucumbers, a variety of sea stars (ochre, slime, pink, sunflower and rainbow), Pacific octopus, spiny dogfish, sea whips, lingcod, perch species, rockfish species and flatfish species.

A large quartz diorite bedrock outcrop showing well developed glacial striations and crescentic gouges indicating southward movement of the ice, lies opposite the park entrance. Note how the outcrop has been polished smooth by the ice.

**Birds**: double-crested and pelagic cormorants, glaucous-winged gull, pigeon guillemot, Oldsquaw. red-throated Loon and three species of grebes.
**Marine Mammals**: Harbour Seal, California sea lion and Steller's sea lion.
**Mammals**: Douglas squirrel, northwestern chipmunk, raccoon, Norway rat, bushy-tailed woodrat, bats and river otters.

### Further Information

To reach the park take Hwy 99 north from Vancouver for 35 km. Visitors may also arrive by B.C. Rail or by private boat. • The campsite in the park has 50 sites plus some walk-in tent sites.

*Barb McGrenere*

# SQUAMISH

Located at the confluence of three rivers and the Squamish Estuary, the area hosts the largest concentration of Bald Eagles in the world (up to 1500) in winter. The 24 km of dykes on the 600 ha of estuary and the nearby Alice Lake Park and Diamond Head Area provide opportunities to search for nearly 200 species of birds reported from the surroundings. The largest numbers of eagles are sighted in the trees along the rivers, particularly on the river dykes near Government Rd in Brackendale, 8 km north of Squamish. Plants are typical fo coastal swamp/rain forest to alpine meadow.

Much of the estuary is owned by the B.C. Railway, but responsible exploration is allowed without seeking permission to enter.

The area is part of the coastal crystalline geological complex. The mountain building is related to the Juan de Fuca plate going under and east in relation to the North American plate being forced up and going west. Stawamus Chief Mountain (el. 651 m) resulted from the interaction of these plates.

**Birds**: (over wintering) trumpeter swan, three loon species and three grebe species, two species of cormorant, surf and white-winged scoters, sea ducks, several gull species, common murre, pigeon guillemot, marbled murrelet, northern pygmy-owl, northern saw-whet owl, mountain chickadee and northern shrike; (summer visitors) three teal species, both yellowlegs, Caspian tern, black and Vaux's swifts, calliope and rufous humming-birds and a wide variety of perching birds; (residents) green-backed heron, large numbers of great blue herons, a variety of ducks, western screech and great horned owls, woodland raptors, red and white-winged crossbills; (migrating) snow goose and an occasional great gray owl.
**Mammals**: river otter, harbour seal, California sea lion, snowshoe hare, Douglas squirrel, black bear, black-tailed deer, raccoon, cougars and mountain goats.

**Further Information**
Squamish lies at the head of Howe Sound, 50 km north of Vancouver on Hwy 99. Rail is an alternative means of transportation. A scheduled passenger train leaves the North Vancouver B.C. Rail Terminal every morning and returns every evening. The trip lasts an hour and 15 minutes and is very scenic. The Public Wharf is available for those coming up by boat. See Chart "3526, Howe Sound" and "Harbour Plans, Chart 3534." • To explore the dykes go to downtown Squamish at 3rd Ave. and Vancouver St. A large sign illustrates the various dyke walks in the estuary. A one-hour stroll on these dykes may be taken by parking at the end of 3rd Ave and Pemberton or further along at the west end of 3rd Ave just before the tracks. To take a short exploratory trip, turn west off the highway at the traffic light on Cleveland Ave. Watch for Buckley Ave heading north.

**Km 0.0:** The corner of Buckley and Cleveland; drive north.
**Km 1.2:** Located at the intersection of Bowen, Buckley and the railway tracks; check the nearby pond or walk around the slew to find waterfowl, including trumpeter swan (November to March).
**Km 1.9:** Turn left just past the feedlot/horse stables. Park near the dyke and walk south for 4 km to the seaward end of the dyke or north approximately 6 km to Brackendale.

**Birds:** bald eagles, trumpeter swans and possibly a peregrine falcon.

**Km 6.7:** At the dyke, opposite Camp Squamish, look for bald eagle.

**Km 8.0:** Turn left on to Judd Rd for 0.9 km to Brennan Rd. Turn left on Brennan and drive another 0.3 km to the end of the road and park under the hydro towers. Walk north on the dyke or south on the sandbars. Watch for green-backed heron, eagles, wood duck and Barrow's goldeneye.

Retrace the route back to the intersection of Judd Rd and Government Rd. Eagles can also be seen north at the Cheakamus River Bridge.

### Further Information

Topographic Maps 92G-11 and 14 should assist with local roads. Other sources of natural history information may be obtained from the Squamish Estuary Conservation Society at Box 1274, Squamish, B.C. V0N-3G0; the Squamish Field Naturalist Club, Box 219, Brackendale, B.C. V0N 1H0.

*Peter Axhorn and Mike McGrenere*

# ALICE LAKE PROVINCIAL PARK

Within an hour's drive of Vancouver, Alice Lake Provincial Park provides access to a typical west coast forest. The park is best in late June to early July when birds and flowers are at their peak.

**Birds**: western screech and northern saw-whet owls, ruffed grouse, American dippers, red-breasted sapsuckers, Steller's jays, gray jays, Townsend's and black-throated gray warblers.

**Mammals**: coast black-tail deer, black bear, raccoon, Douglas red squirrel, varying hare and flying squirrel.

**Amphibians**: red-legged frog, Pacific tree frog, northwestern toad, alligator lizard and several salamanders.

**Fish**: rainbow and cutthroat trout and dolly varden and Splake (a cross between lake trout and brook trout).

**Plants**: Douglas-fir, Western red-cedar, Western hemlock, vine maple, devil's club, salal, wild ginger, liverworts, sundew, buckbean, bog cranberry, swamp laurel, white and yellow water lilies, pink corydalis, slime molds and mushrooms.

### Further Information

Take Hwy 99 north from Vancouver to beyond Squamish and watch for the signs to the park. A park interpreter is stationed at the park during the summer. A map of the park and a brochure for the Swamp Lantern Intrepretive Trail may be obtained at the park headquarters north of the campground and public areas.

*Barb McGrenere*

# BIRKENHEAD LAKE PROVINCIAL PARK

Annual rainfall at Birkenhead is about one-third that of Squamish and hence the park lies between the coastal wet forest and the drier interior country. Douglas-fir of the west mixes with ponderosa pine of the interior. Similarly the birds to be found are a mix of the coast and the interior.

The Kokanee, a small landlocked salmon, spawns in Phelix Creek at the north end of the park each fall.

**Birds**: red-breasted and red-naped sapsuckers, Vaux's swift, Hammond's flycatcher, gray and Steller's jays, chestnut-backed chickadee, Townsend's and MacGillivray's warblers, red crossbill, blue grouse, Townsend's solitaire, great gray owl, three-toed woodpecker, white-tailed ptarmigan and rosy finch.
**Mammals**: black bears, pine martens, bobcats, river otters, mountain goats, moose and mule deer. Smaller mammals that are more likely to be noted include snowshoe hare, northwestern chipmunk, Douglas squirrel and little brown bat.

**Further Information**
Drive north from Vancouver on Hwy 99 to Pemberton, which is located north of Garibaldi Provincial Park. Continue on north from Pemberton to Mt Currie and stay on the paved road to D'Arcy. Just before D'Arcy, at Devine, (37 km northeast of Pemberton), turn west onto the unpaved secondary road and proceed 18 km to the park entrance. • There is a campground in the park.

*Mike McGrenere*

# PEMBERTON VALLEY

The Pemberton Valley with its fertile soils plays host to the western limits of several "interior" bird species.

Lying north of Garibaldi Provincial Park, the valley is the nearest place a Vancouverite can travel to see the more eastern species of plants and birds.

**Birds**: Wilson's, orange-crowned, MacGillivray's and black-throated gray warblers, eastern and western kingbirds, American redstart, Cassin's finch, veery, lazuli bunting, northern oriole, red-naped and red-breasted sapsuckers, black-headed grosbeak and gray catbird.
**Mammals**: moose, black-tailed deer, black bear, coyote, bobcat, mink, marten and short-tailed weasel.
**Fish**: rainbow and cutthroat trout, coho, sockeye and chinook salmon.
**Plants**: cottonwoods, Douglas-fir, Western red-cedar and western hemlock.

**Further Information**
The Pemberton valley is reached by car from Vancouver on Hwy 99. From Lillooet, drive the Duffey Lake Rd which is passable only in summer. The B.C. Railway has passenger rail service from Vancouver.

*Mike McGrenere*

# JOFFRE LAKES / PEAK AREA

The Joffre Lakes area is one of the finest sub-alpine areas in the Coastal Mountains. Three lakes, avalanche paths, alpine meadows and the 3000 m Joffre Peak are there to explore. The 5-km hike with a 397-metre altitude gain, takes at least 6 hours for the round trip hike.

Glacial activity near the upper lake has produced a "textbook" example of a lateral moraine.

**Birds**: mountain, and chestnut-backed chickadees, red and white-winged crossbills.
**Mammals**: Pikas.
**Plants**: (forest) alpine fir, mountain hemlock, western hemlock, Englemann spruce, Canada blueberry, false azalea, copper bush, trailing rubus, queen's cup, bunchberry and oak fern; (bog) lodgepole pine, Labrador tea, alpine swamp laurel and cranberry; (mountain) shrub alder, devil's club, stink currant, lady fern, Indian hellebore, salmonberry, foam flower, twisted stalk, mountain valerian, oak fern, alpine fir and whitebark pine, stunted mountain hemlock, dwarf juniper, dwarf willow, dwarf huckleberry and white moss heather,azalea, meadow spirea (primary ground cover), mountain sorrel, fringed grass of parnassus and alpine fireweed.

**Further Information**
Drive north from Vancouver on Hwy 99 to Pemberton and then on to Mt Currie. Continue east on the Duffey Lake Rd (summer use only and refer elsewhere) to the trailhead. The parking lot at the start of the trail is about 35 km east of Pemberton. The hike ends near the foot of the glacier along the shores of the alpine lake. Climbers continue on to Joffre Peak and Mt Matier.

*Mike McGrenere and Gerry Hallett*

# PEMBERTON TO LILLOOET VIA THE DUFFEY LAKE ROAD

From mountain goats to the largest collection of Indian pictographs in B.C. can be seen on this short cut across the mountains.

**Km 0.0:** Pemberton junction (intersection of Hwy 99 and the Pemberton-Mt. Currie Rd.

**Km 7.1:** Junction, a right turn (east) will lead to Lillooet Lake and the Duffey Lake Rd. Straight ahead leads to D'Arcy and Birkenhead Lake Provincial Parks.

**Km 16.2**: Lillooet Lake and the Lillooet River delta. This delta is one of the fastest growing deltas in the world.

**Km 18:** Turn left on to the Duffey Lake Rd. This road climbs quickly up the Joffre Creek valley and is not suitable for trailers.

**Km 22:** Watch for Mule Deer that are often sighted in this logged off area as they browse on the young shoots of the second growth.

**Km 29.9:** Side road on the right, keep left.

**Km 30.6:** Dividing point between the coastal and interior watersheds.
**Km 43.1**: Junction, continue straight or left. Duffey Lake on the left is very cold since it is covered with ice for much of the year.
**Km 49.9:** Duffey Lake narrows to become Cayoosh Creek.
**Km 77.2:** Watch for Mountain Goats in this area.
**Km 91.9:** Road Junction, continue straight ahead.
**Km 94.6:** Bridge over Cayoosh Creek. The largest collection of Indian rock paintings found in B.C. lies across the creek on the rock bluffs above the railway tracks. To visit the site, turn left just beyond the bridge and drive into the Evans Logging Company yard and ask permission to cross the property and park your car. Scramble up to the railroad tracks and walk along these tracks for about five minutes. Look for a narrow trail leading up a steep slope to a huge overhanging rock face with a large concentration of pictographs.
**Km 99.6:** Mile zero cairn in Lillooet is built of rocks left over from placer mines.

**Further Information**

To reach this overland route, take Hwy 99 north from Vancouver to Pemberton. There are no gas stations or other services along this road. Travellers are advised to have a full can of gas and carry necessary emergency equipment on the trip. • *Lower Mainland Backroads: Volume 1 - Garibaldi to Lillooet* by Richard and Rochelle Wright, 1977. Saltaire Publishing Ltd., Sidney, B.C.

*Barb and Mike McGrenere*

# SOUTHERN CHILCOTIN MOUNTAINS

These wilderness valleys and mountains lie within five hours driving time from downtown Vancouver. Wildlife is plentiful in these valleys and mountains.

It takes a week to properly explore a part of this area. Hiking or a horse trip are the best ways to see it.

The area lies in a transitional habitat from the rugged coastal mountains of the west to the dry interior plateaus in the east. The bird species to be found reflects this range of habitats.

The geology of the Chilcotins provides an interesting diversity of formations, including coastal granites, volcanic basalts and mineralization. Highly fossilized deposits are also found. Castle Peak in the north central area, consists of outcrops containing up to 90% ammonites and trilobites. Most of the other peaks are erosional remnants of volcanic rock extruded 30-40 million years ago. Some of these peaks remained above the ice during the recent glaciation and hence served as a refugium for vegetation. Three plants found on these peaks — a poppy, fleabane and buttercup — are not found again for 350-400 km to the south.

**Birds**: spruce and ruffed grouse, bald and golden eagles and white-tailed ptarmigan great gray owl, meadow lark, blue and ruffed grouse, prairie falcon, white-crowned and golden-crowned sparrows and Vaux's swift.
**Mammals**: mule deer, black and grizzly bear, hoary marmot, coyote, California bighorn sheep, mountain goat, moose, wolverine and wolf.
**Fish**:Rainbow trout, dolly varden
**Plants**: bluebunch wheatgrass, balsamroot, spring beauty and yellow avalanche lily.

**Further Information**
The nearest community, Gold Bridge, may be reached year-round via the Fraser Canyon to Lillooet and then west, or by the Squamish Hwy via Pemberton and then north on a couple of routes depending on snow conditions and time of the year. Travel time on either route is about 5 hours from Vancouver. Accommodation is available in Gold Bridge. Forest Service campsites are located throughout the area. • Maps include: 1:50,000 Tyaughton Creek 92-0/2; Warner Pass 92-0/3. The Forest Service Map is Spruce Lake Recreation Map.

*Gail Ross*

# PINE MOUNTAIN ECOLOGICAL RESERVE

One of the last peat moss-sedge bogs remaining in the Lower Mainland, it offers opportunities to find a wide range of plants, sandhill cranes in summer, and, in winter northern shrike, rough-legged and red-tailed hawks

Pine Mountain is a rocky outcrop in the middle of a bog. Nearly 130 species of birds have been spotted in and around the site. Over 200 different flowering plants have been identified on the site, with vegetation ranging from wet bog to higher rocky outcrops.

**Birds**: sandhill cranes, Vaux'swifts, barn, western screech and long-eared owls, Black-throated gray warbler, black-headed grosbeak, Trumpeter swan and northern shrike.
**Mammals**: black bear, black-tailed deer, River otter and beaver.
**Plants**: cloudberry, round leaf sundew, heart-leaved twayblade cordata and rusty saxifrage.

**Further Information**
Take Hwy 7 (Lougheed Hwy) east from Vancouver and cross the Pitt River Bridge; immediately over the bridge watch for Dewdney Trunk Rd on the left (north) side, going straight east at an angle off the highway. Go east on Dewdney until you come to a "T" junction; turn right, go approximately100 metres to a 3-way stop sign; turn left and drive to Neaves Rd on the left. Take Neaves Rd north over two branches of the Alouette River. As you drive north the name "Neaves" changes to Rannie Rd. A short drive north on Rannie cross Sturgeon Slough, then over a gravel road where hydro pylons cross. A few metres past the gravel road, alongside the pine trees, watch for a small gravel pull-off on the right (east) side of Rannie Rd. A plank rests across the ditch to provide access into the reserve. The trail winds through the bog and up to the top of the rock.

*Duanne van den Berg*

# UNIVERSITY OF BRITISH COLUMBIA RESEARCH FOREST

This well-managed forest contains a great variety of natural history features including abundant mushrooms among the mature conifers and in the marginal areas. With six lakes and diverse forests, the area abounds in birds. Lying immediately west of Golden Ears Provincial Park this research station, owned and managed by the University of B.C. and containing 42 square km, is well within an hour of downtown Vancouver.

The forest complex provides excellent habitat for birds along the margins. Species typical of the coast mountain region are common. The site is good for saprophytic plants including coral root and Indian pipe.

**Further Information**
To reach the site, take Hwy 7 (Lougheed Hwy) east from Vancouver to the Dewdney Trunk Rd west of Maple Ridge; turn off east on Dewdney and follow directional signs for Golden Ears Provincial Park; turn north on 232. Continue north past the Golden Ears turn off until 232 becomes Silver Valley Rd. and follow it to the end into the research station. • Camping is available in Golden Ears Provincial Park.

*Peter Perrin*

# GOLDEN EARS PROVINCIAL PARK

Golden Ears contains the most complete example of the Coast Mountains in the provincial park system, and abuts the southern boundary of Garibaldi Provincial Park. Over 140 species of birds have been reported from the park.

Spirea Bog Interpretive Trail, located on the east side of the park road, north of Park Headquarters and south of Alouette Lake, is an ideal site to explore bog natural history. Sphagnum, a species of sundew, kalmia, Labrador tea and several other plants that have adapted to wet, acidic conditions make this site worth stopping to examine.

**Birds**: Osprey, red-breasted sapsucker, warbling vireo, black swift, Vaux's swift, western screech, pygmy and barred owls, red-breasted sapsucker, five species of swallow, bushtit, winter and Bewick's wren, Hutton's vireo, black-throated gray warbler and black-headed grosbeak.
**Mammals**: Mountain goat, black-tailed deer, beaver, black bear and Douglas squirrel.
**Plants**: yew trees, water hypericum, round leaf sundew, Indian pipe, gnome plant, pinesap, marsh violet and skunk cabbage.

**Further Information**
Take Hwy 7 east from Vancouver for 41 km to Maple Ridge and the park sign. Follow these signs northeast into the park. A 343-unit campground is on site. The Mike Lake Trail which circles the lake is the best place for nature observation. For more information contact the park at P.O. Box 7000, Maple Ridge, B.C. V2X 7G3.

*Gail Ross and Al Grass*

# HARRISON MILLS NICOMEN SLOUGH

A hundred bald eagles soaring against the fir-clad mountains while hundreds more are clustered on the huge cottonwoods is one of North America's great wildlife spectacles.

Every winter the eagles gather on the lower reaches of the Harrison River to feed on spawned-out Pacific salmon. A winter flock of trumpeter swans can be watched at close range, west of Harrison Mills. Both eagles and swans are here from November to March with the best dates from December to February.

**Birds**: bald eagle, trumpeter swans, ducks, ravens, northwestern crows, dippers, Steller's jay, chestnut-backed chickadee and varied thrush.

**Further Information**
Harrison Mills lies on the north side of the Fraser River, 100 km east of Vancouver, or 21 km west of Harrison Hot Springs, along Hwy 7. A dyke prevents viewing Nicomen Slough from the highway, but there are several places where a vehicle may be parked safely. Drive 14 km west of Harrison Bridge and park. Walk to the top of the dyke and watch the wintering flock of trumpeters at close range.

*David Stirling*

# CULTUS LAKE PROVINCIAL PARK

The park is known for its relatively warm lake. Phantom orchids and Pacific giant salamanders, both endangered species and protected, have been found here.

Over 93 species of birds have been recorded within the park. The second growth mature maple forest that characterizes much of the park provides habitat for a number of deciduous-loving birds not found in abundance elsewhere in the Fraser valley.

Some very large firs, over 100 m high and bearing fire scars, remain from a once extensive forest of climax Douglas-fir.

**Birds**: red-eyed vireo, black-throated gray warbler, pileated woodpecker, pygmy and western screech owls, black and Vaux's swift, red-breasted sapsucker, bushtit, winter and Bewick's wren, ospreys, Canada geese, mallards, goldeneyes and buffleheads.
**Mammals**: shrew, blacktail deer, coyote and shrew-mole.

**Further Information**
The park lies 10 km off the Trans Canada Hwy via Sardis and 16 km via Yarrow. The Columbia Valley Hwy, an interesting drive through farms and woodlots, bisects the southeast section of the park.

*Gail Ross*

# AGASSIZ - MARIS SLOUGH

Flocks of up to 100 trumpeter swans may be found from late October to early February, together with a wide collection of other wintering birds, plus river otter, on this free-flowing arm of the Fraser River. All times of the year are rewarding for naturalists in this area of the lower mainland. Lying an hour and a half drive east of Vancouver this site offers marsh, water, deciduous and coniferous woods and farmland to explore in a half day trip.

The Seabird Island Indian Band installed a spawning channel on the reserve east of the intersection of Seabird Island and Chaplin Rds in 1984. A weir trap is in operation on Chaplin Rd arm of the slough during September to early December. Dipper frequently visit the trap during its operation, presumably in search of salmon eggs. A belted kingfisher also visits the weir trap, in search of bullheads.

**Birds**: bald eagle, grebes (pied-billed and the odd red-necked), mergansers(common and hooded), dipper, northern shrike, osprey, green-backed heron, Steller's jay, a wide variety of song and perching birds and turkey vulture.
**Mammals**: river otters, beavers.
**Fish**: Pink, chum and sockeye salmon (late fall to March), Bullheads, trout, carp.
**Plants**: willow, alder, dogwood, cottonwood, fir, grass, shrubs, big leaf maple, black cottonwood, western hemlock, Douglas-fir and Western red-cedar.

**Further Information**
Drive east from Vancouver on the No. 1 Fwy to the junction with Hwy 9. Continue on Hwy 9 to 1 km east of Agassiz and turn north on Seabird Rd. Follow this road north east to Chaplin Rd and follow Chaplin around the north east arm of the slough. Chaplin Rd dead ends so retrace your route back to Seabird Island Rd, turn east and follow the south east arm of the slough through the Indian Reserve back to the junction of Waleach Rd and the No 9 (Haig) Hwy.

*J. Janne Perrin*

# SASQUATCH PROVINCIAL PARK

Legends talk of large, hairy, man-beasts of the woods that gather every nine years in Morris Valley, only minutes away from Harrison Hot Springs. Other stories tell of tall, dark Bigfoot prowling about villages, ransacking ladders and disappearing noisily into dense forest. Hence the name.

Lakeshore Paths have been built at both Deer and Hicks Lake campgrounds. An additional 5 km trail and boardwalk around Hicks Lake provides a route under broad-leaved and vine maple canopies, through second-growth hemlock and Douglas-fir forest and past salmon berry and fir lined streams. Round-leaved sundews, Indian pipe, maidenhair fern, and ostrich fern are some special plants,.

**Birds**: yellowthroat, swift, bufflehead, pied-billed grebe, ospreys. common loon, spotted sandpiper, woodpeckers, barred owls, merganser, red-necked and pied-billed grebes, both scaup species, great blue heron and black-throated gray warblers.

**Mammals**: pika, beaver and mountain goats.
**Amphibians**: Alligator lizards, Red-legged frogs, Pacific tree frogs and salamanders.
**Fish**: three-spine stickleback (Deer Lake), White sturgeon and longfin (Harrison Lake), Rainbow and cuthroat trout (all three park lakes).

**Further Information**
Take Hwy 7 east of Chilliwack to Harrison Hot Springs and then drive 12 km north to the park.

*Gail Ross*

# SKAGIT VALLEY

Straddling the transition between coastal and interior forest zones, this broad U-shaped valley is quite unique in the Lower Mainland area and has many distinct plant communities within a small area. The two large meadows in the south part of the valley both contain an intermingling of coast and interior plants and are full of wildflowers in spring and early summer. The valley is an important spring migration route for many bird species including the western bluebird. Over 200 bird species have been recorded in the area. This is the western limit for Englemann spruce and ponderosa pine; and the eastern limit for broadleaf maple, amabilis fir, yellow cedar (except for an isolated population in Valhalla Provincial Park). The Skagit is known for stands of Pacific rhododendron which extend up the river from Washington State to the roadside display in Manning Park. The river is the most productive stream fishery in the Lower Mainland.

**Birds**: long-billed curlew, lesser yellowlegs, western bluebird, poorwill, ruffed, blue, spruce grouse, American crow, northwestern crow and yellow-bellied and red-breasted sapsuckers.
**Mammals**: mule deer (spring densities are among the highest in the region), bear, beaver, cougar, coyote, bobcat, river otter, mink, marten, raccoon, squirrel, chipmunk,snowshoe hare, elk, moose (Ross Lake area), mountain goats (high slopes of Silvertip).
**Reptiles and Amphibians**: three species of garter snake, rubber boa, northern alligator lizard,two species of salamander, three species of toad, red-legged frog.
**Fish**: Rainbow trout, Dolly Varden, eastern brook trout, coastal cutthroat trout.
**Trees**: Douglas-fir, Western red-cedar, western hemlock, lodgepole pine, black cottonwood, trembling aspen, Pacific rhododendron, ponderosa pine, water birch and Pacific crab apple.
**Plants**: Mountain juniper, small flowered alumroot and rosy pussytoes.

**Further Information**
To drive to the valley take the turnoff from Hwy 1, to the east, 2 km west of Hope. The 60 km gravel road extends south to the U.S.A. border and beyond Ross Lake in the North Cascade National Park complex in the U.S.A. Suggested reading: Perry, Thomas L. *A Citizen's Guide to the Skagit Valley.* • Additional reports are available from the B.C. Parks in Victoria.

*Gail Ross*

# HOPE TO PRINCETON — HWY 3

**Km 8.0:** Nicolum Provincial Park has good birding in April.

**Km 17.5:** The Hope Slide. Mountain goats can often be seen on the rock face to the south of the slide.

**Km 18.5:** Tashme Marsh on the north side of the highway, offers good viewing opportunities for many species of water and marsh birds.

**Km 19.0:** Now a recreational development, Sunshine Valley was originally the site of the vanished town of Tashme. At the doutbreak of World War 2, the Canadian government relocated 2,300 people of Japanese origin from coastal B.C. to a ranch owned by A.B. Trites. For the next 4 years, the men worked about the camp and on the Hope-Princeton Highway. At the end of the war the government sent crews in to tear the town apart.

**Km 26.0:** A hoary marmot on the sign marks the west gate of Manning Provincial Park. These marmots can be seen in summer in Blackwell Meadows, accessible along a 15 km paved and gravel road departing north off the highway, from Manning Park Lodge. At the park entrance, to the north of the parking lot, a short 20-minute return loop walk leads to an impressive section of stone-work road built by the Royal engineers in 1860. The trail to Mt. Outram and Ghost Pass Lake also begins at this parking lot.

**Km 30.5:** Another section of the Royal Engineer's road that was built as a wagon road to replace the first 40 km of the Dewdney Trail. Along this section of the highway, willow, red osier dogwood, red alder, Western red-cedar and western hemlock flourish. The towering cliffs across the river are home to a small herd of mountain goats.

**Km 37.5:** Rhododendron Flats, only 2 km further along from the above site, has a totally different combination of plants.

**Km 39.0:** At Snass Creek, the Dewdney and Whatcom trails follow the same route and veer north toward Paradise Valley and the Punchbowl. This route provides opportunities in May to find trillium, ginger, and calypso orchid and to listen to the song of the winter wren in the forests when other higher trails are snowbound.

**Km 44.0:** On Cayuse Flats, in fall, the black cottonwood stands provide a yellow border along the road. On the rocky slope above, Rocky Mountain juniper and Douglas-fir thrive in areas with good drainage and sunlight.

**Km 50.0:** This large area of sun-bleached snags and young growth is the result of a fire in 1945. Deer thrive on this new growth and bears harvest the prolific crops of blueberries and huckleberries. Blue and spruce grouse use the fallen logs as hooting perches, while red-tailed hawks can often be seen soaring overhead.

**Km 69.0:** The area around Manning Park Lodge is a good spot to find mountain chickadee, raven, cowbird, swallows and kinglets. The predominant tree in this area is the lodgepole pine.

**Km 70.5**: Beaver Pond; A variety of birds are found here in May or June.

**Km 84.6:** As you drive east, the climate becomes increasingly drier. Here at McDiarmid Meadows look for aspen, ponderosa pine, Jacob's ladder and stonecrop.
The meadows around the old homestead are a favoured location for Macgillivray's warbler, American kestrel, common nighthawk, coyote and mule deer.

**Km 123.0**: These ponderosa pine occur in the B.C. Interior in the zone of low rainfall. Birds to watch for include mountain bluebird, especially in the nest boxes along the highway further east, and pygmy nuthatch. Hawks often soar by here.

**Km 137.5:** Princeton

**Further Information**

Hope is the western terminus of Hwy 3, where it intersects with Hwy 1 and the Coquihallo, Hwy 5. Princeton is the south end of Hwy 5A, south of Kamloops.

*Gail Ross*

# MANNING PROVINCIAL PARK

Straddling Hwy 3 for 64 km, Manning Park offers a tremendously rich and diverse landscape. Lying between the coast and the interior, the park encompasses a variety of biogeoclimatic zones. They range from coastal western hemlock forests to the dry interior Douglas-fir forests. Englemann spruce-subalpine fir zone lies in the central part of the park. There is high mountain vegetation (alpine tundra), too. This diversity has produced over 700 flowering plant species within the park. It is the northern limit of the red rhododendron and the western limit of alpine larch and whitebark pine.

About 200 bird species have been recorded in the park. A check list is available at the Visitor Centre. The park is known for its butterflies, including indra swallowtail, green-veined white, Compton's tortoise-shell, green hairstreak, blue copper, silvery blue and the oregonian.

Two geological faults transect the park, the Pasayten on the east, and the Hozameen to the west. These faults have moved different rock types laterally so that today they sit side by side. The land between these two faults has been dropped downward to form a graben. This trough has protected about 7300 m of relatively young (75-135 million years old) Cretaceous rocks from erosion. These layers have been removed in all other places in the Cascade Mountains.

**Mammals**: mule deer, black bear, hoary and yellow-bellied marmot, Columbian and golden mantled ground squirrel, Townsend's and yellow-pine chipmunk, Douglas and red squirrel, snowshoe hare, lynx, river otter and marten.

## Sumallo Grove

Located at the junction of the Skagit and Sumallo River, this grove of tall trees, including Douglas-fir, Western red-cedar and western hemlock, should be strolled through for a full appreciation of their height. An interpretive trail through this towering forest can be enjoyed. The Skagit River trail begins at this grove and heads south.

**Birds**: common yellowthroat, western flycatcher, yellow-rumped warbler, harlequin duck, swifts and winter wren.
**Plants**: calypso orchid, trillium, ginger, willow and red osier dogwood.

## Rhododendron Flats

Nine coniferous species grow in this small area, including western white pine and three species of true fir (*Abies*). The main attraction of this spot is the annual flowering of the Pacific rhododendron that usually peaks in early June. This relatively large colony is the most northerly occurrence of the shrub in B.C.

The spot is a good place to search for western teaberry and several saprophytic species including pinesap and Indian pipe. In summer Townsend's warblers are usually abundant in the trees overhead.

## Canyon Nature Trail

Located just south of the Manning Park Lodge along the Lightning Lake Rd, this trail provides a good hour of nature exploring. Interpretive brochures are available.

## Strawberry Flats

Lying south of the Lodge beyond Lightning Lake, the flats are known for their incredible floral diversity with over 150 flowering plants to be found at the meeting of coastal, interior and alpine habitats. Mid-July is peak flowering time. The combination of floral abundance, forest and fire succession, the profusion of wild strawberries and subalpine meadow draws a variety of birds and butterflies.

**Birds**: chestnut-backed, mountain and boreal chickadees, pine grosbeak, rufous and calliope hummingbirds, and spruce grouse.
**Mammals:** black bears.

## Lightning Lakes

From the Lightning Lake campground south of the Manning Park Lodge, there is a loop trail around the lake or continue hiking southwest and back along Lightning Creek.

**Birds**: three-toed woodpecker, common loon, Townsend's warbler, goldeneye duck, shorebirds, bald eagle and osprey.

## Beaver Pond

The site has Lincoln's sparrow (breeds here), belted kingfisher, rough-winged swallow and sora. Other birds that nest here in May and June, include several warblers, four species of flycatchers, tree swallow, rufous hummingbird and spotted sandpiper. Mammals to watch for include muskrat, beaver, the rare moose and yellow-pine chipmunk. Western spotted frog and dragonflies are common. A brochure is available for the interpretive trail.

### Frosty Mountain

The parking lot and trail lie just west of the Beaver Pond. The highest (2408 m) peak in the park offers alpine vegetation as well as the largest and most westerly grove of alpine larch in the province. In the high country look for rosy finch and white-tailed ptarmigan. On the way up watch for spruce grouse.

North of the highway, lodge and main administration area, a road goes up to Cascade Lookout. From here you can hike up to some great nature spots including the next two sites.

### Dry Ridge Trail

The ridge is home to plants that are native to the lowlands of the dry interior, as well as plants from the alpine tundra. Very few subalpine species flourish on the ridge, and thus the transition of plants from valley bottom to mountain top is absent.

### Blackwall / Three Brothers Subalpine Meadows

These meadows are accessible by gravel road and are some of the most extensive in the province. A self-guiding brochure for the Paintbrush Nature Trail is available during the summer months. The meadows are in peak bloom from mid-July to early August. Look for alpine larch and whitebark pine which are at the western limit of their range. Locate the nearby colony of hoary marmots. Interpretive walks are given in the summer months.

**Further Information**

Hwy 3, east of Hope off the Trans Canada, runs through the middle of the park. Manning Park Lodge Resort is located in the central core of the park and offers rooms, cabins, chalets and a restaurant. There are four provincial campgrounds open from May to October. A winter campground is located at Lone Duck Bay southwest of the Visitor Centre. The Greyhound Bus has a stop at Manning Park Lodge. • Gas is available on the east and west boundary. Telephones are found at Manning Park Resort and at both the east and west park entrances. The nearest doctors and/or Hospitals are Hope (65 km from the Resort) and Princeton (about 67 km from the Resort). • Suggested maps include: Topographic map 92/H; Ministry of Environment Map of Manning Park (1:50,000); Outdoor Recreation Council Map #8 (Princeton-Manning-Cathedral, 1:100,000).

**Suggested Reading:**

Coates, J.A. *Geology of the Manning Park Area.*. Geological Survey of Canada. Bull. No.238, 1974.
Grass, A. and G. Ross. "Land of the Golden Larch." *Discovery*, Summer 1984.
Harcombe, A. and R. Cyca. *Exploring Manning Park.*. Douglas and McIntyre, Vancouver, 1979.
Ross, G. "The Rhododendron Roadside Attraction," *Nature Canada*, April 1983.
Ross, G. and A. Grass. "Where There's Wildlife." A brief guide to the birds and mammals of Manning Park.. *Discovery*, summer 1983.
Provincial Park reports are available at the park office or from headquarters in Victoria.

*Gail Ross*

# CASCADE WILDERNESS

Lying adjacent to the northwest portion of Manning Park, the Cascades present an opportunity for hikers to step back 100 years and experience the natural history of an area almost untouched by man. The Indian trails, fur trade paths and later routes made by those on their way to search for gold are still there to experience.

Situated within the North Cascade Mountains, this wilderness area lies just east of Hope and west of Manning Park.

## Dewdney and Whatcom Trails (Snass Creek)

A hiking loop may be taken north on the Dewdney Trail for about 15 km and then back south via the Whatcom trail. The hike begins by passing through a beautiful stand of red-cedar, Douglas-fir and western hemlock. Birds typical of old-growth forests can be encountered here. The forest floor is decorated with calypso orchids, bunchberry and wild ginger. Natural openings caused by talus slopes and avalanches at the upper ends of valleys provide habitat for a variety of wildlife not found in the forests, including marmots and pika. These clearings often contain stands of stinging nettle coupled with cow parsnip and slide alder. Above Dry Lake, Snass Creek disappears underground.

At Paradise Valley, near the top, the area includes the headwaters of the Tulameen River. The upper valley is representative of Englemann spruce-subalpine fir associations and offers a high elevation, cascade valley-meadow landscape full of flowers in mid to late July.

The return trip on the Whatcom Trail climbs up through an Englemann spruce-subalpine fir forest to Punch Bowl Lake and the Cascade Divide. Vegetation at the divide consists of a wonderful combination of mountain hemlock, amabilis fir, subalpine fir, and Englemann spruce. The trail then passes through meadows, avalanche shutes and out to the coastal forest, Snass River and parking lot.

**Further Information**

The parking lot and trailhead are located on Hwy 3 just north of Rhododendron Flats to the east of the west gate of Manning Provincial Park. Begin on the Dewdney Trail and hike north east for 15 km to Paradise Valley, where the trail rejoins the Whatcom Trail. The Whatcom Trail returns south for 17 km via Punch Bowl Lake to the parking lot. • Topgraphic Maps include: 1:50,000, 92 H 7 and 92 H 6 (east half). • An excellent guide to the numerous trails of the area, *Old Pack Trails in the proposed Cascade Wilderness* by R.C. Harris and H.R. Hatfield, is available from the Okanagan Similkameen Parks Society, Box 787, Summerland, B.C. V0H-1Z0.

*Gail Ross*

# SOUTH AND SOUTHEAST BRITISH COLUMBIA

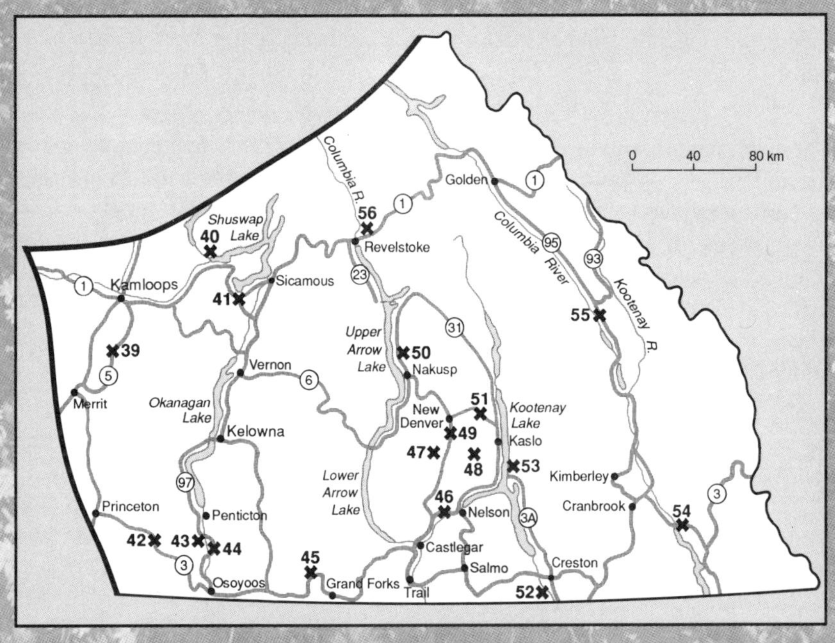


39. Kamloops to Merrit - Hwy 5 and Side Trips
40. Adams River Area
41. Salmon Arm
42. Princeton to Osoyoos - Hwy 3
43. Vaseux Lake
44. Osoyoos North to Penticton - Hwy 97
45. Nancy Green P.P. via Grand Forks to Rock Creek - Hwy 6
46. Castlegar via Nelson to Salmo - Hwy 3A and 6
47. Valhalla P.P.
48. Kokanee Glacier P.P.
49. South Slocan via New Denver to Nakusp - Hwy 6
50. Arrow Lakes - Fauquier to Galena Bay - Hwys 6 and 23
51. New Denver to Kaslo - Hwy 31A and North Hwy 31
52. Creston Valley Wildlife Area
53. Kootenay Lake Area
54. Alberta Border - Cranbrook - Kimberley - Creston - Hwy 3
55. Golden - Radium - Cranbrook - Hwys 95 and 93/95
56. Golden - Glacier - Mt Revelstoke - Sicamous - Hwy 1

# *KAMLOOPS TO MERRITT — HWY 5A AND SIDE TRIPS*

The prairie-like Knutsford Plateau from Kamloops south to Nicola Lake and the interspersed wetlands provide opportunities to spot hawks and a variety of upland and water birds, coyotes, deer and numerous spring and summer flowers.

Extensive flooded areas in the spring attract thousands of migrating waterfowl and shorebirds which use the area as a staging location on their way into the north.

Vegetation on the plateau is primarily grassland giving way to forest at approximately 900 metres. Big sagebrush is the dominant shrub in the lower grasslands. Ponderosa pine is the main tree.

These highlands, known as the Kamloops Midlands, extend from Knutsford in the north to Nicola Lake near Merritt. They slope towards the Thompson River and are thought to be a remnant of an ancient valley. During the last glaciation of 19,000 to 10,500 years ago, ice moved in a southwesterly direction over the midlands, scouring out numerous potholes. As the ice melted, the resultant deposits included drumlins and morainal features. Good examples of such glacial action are seen on the east side of the road at Km 7.8 south of Kamloops and looking north from Shumway lake at Km 15.8.

Much of the Thompson-Okanagan area was covered with volcanic-related rocks in the early-middle Tertiary era. These volcanics are well exposed at Km 33.2 south of Kamloops and at the south end of Napier Lake at Km 34.4. This latter outcrop acted as a dam for glacial Lake Merritt and prevented any further erosion by the meltwater outflow.

**Birds**: horned lark, vesper and savannah sparrow, mountain bluebird, long-billed curlew, western meadowlark, cliff and barn swallow, kestrel, red-tailed and Swainson's hawks, rough-legged hawk (October - April); peregrine and prairie falcon, snowy owl (December - February); short-eared, great horned and pygmy owls, hooded merganser, red-necked and pied-billed grebes, sora, marsh wren, Brewer's and red-winged blackbirds, lesser scaup, mallard, blue-winged and cinnamon teals, Ospreys, bald eagle, golden eagle, marsh hawk and Sandhill crane.
**Mammals**: yellow-bellied marmot, coyotes, black bear, mule deer, moose, marten, fisher, mink, snowshoe hare and lynx.
**Plants**: bluebunch wheat grass, needle and thread grass, Sandberg's bluegrass, pasture sage, pussy toes and rabbit-bush (grassland); pink pyrola, one-sided pyrola, bearberry, twinflower, princes pine; (forest) yellow water lily, floating leafed pondweed, duckweed, northern water milfoil, sedge, hardstem bulrush, cattail, spike rush and horsetail (wetland).

## KAMLOOPS TO MERRITT

**Km 0.0:** Begin at the junction of Hwy 1 and Hwy 5A in Kamloops. As you climb the hill heading south, you are entering the prairie-like Knutsford Plateau. Big sagebrush is the dominant shrub for the first 2 km with a variety of grasses (see above).

**Km 6.1:** Turn left or east onto Rose Hill or Old Rose Hill Rd for a good side trip to explore for birds.

**Km 9.8:** Separation Lake is one of the few permanent wetlands on the plateau. The lake is a main migration and staging water body for waterfowl. The uplands are heavily grazed by cattle, restricting nest sites for waterfowl. The lake is rich in invertebrates including amphipod (shrimp) and gastropods (snails). As soon as the ice goes out, usually in late March, numbers of canvasback, redhead, pintail, wigeon, greater and lesser scaup, mallard, tundra swan and Canada geese arrive. Then come rafts of coot, ruddy duck, Barrow's goldeneye and red-necked grebe. Old squaw, surf and white-winged scoter are occasionally observed in spring migration.

At the northern end of Separation Lake, killdeer, spotted and solitary sandpipers, long-billed dowitcher, pectoral, least, and Baird's sandpipers and both species of scaup are commonly seen on spring and fall migration. Others such as black-bellied plover are occasionally noted.

**Km 10.8:** Immediately south of Separation Lake on the east side of the road, a small pond is host to numerous waterfowl especially coot, ruddy duck and eared grebe which nest in a loose colony. Another pond, directly across the road, is a good breeding site for lesser scaup, ring-necked duck, ruddy duck, redhead, coot, sora and horned grebe.

**Km 11.5:** The wild hay meadow on the west side of the road is usually flooded in spring providing a haven for green-winged, cinnamon and blue-winged teal, mallard, pintail, wigeon, killdeer and yellowlegs.

**Km 12.7:** The wetlands, on the east side of the highway, are thick with bulrush. Look for yellow-headed and red-winged blackbird, sora and marsh wren.

Continuing south along Hwy 5A, below the Knutsford Plateau, you reach a series of lakes including Shumway, Trapp, Richie and Napier which lie in a meltwater channel formed at the end of the last ice age. This channel drained north from glacial Lake Merritt and flooded the present Stump Lake, Nicola Lake and Douglas Lake valleys. Waters were held by an ice dam blocking off the Nicola River near Lower Nicola. When the dam melted, the sediments that had built up closed off this meltwater channel forming the present lakes. Look for the drumlinoid ridges which were built up by the southward flowing ice. These ridges occur on the side hills at Shumway Lake and continue down the corridor further south.

The south end of Shumway Lake is a productive cattail marsh and a good place to observe a variety of breeding waterbirds.

**Km 23.5:** The Roche Lake turnoff is also the outlet of Trapp Lake. Although very small, this is an extremely productive marsh with a bullrush perimetre. The water is rich in invertebrates. In spring the ice at the outlet melts early attracting a wide variety of water birds. From June to early September look for numerous breeding waterbirds. Bulrushes are host to a large number of nesting red necked grebes.

Ospreys (April to September) and bald eagles (fall and spring) can be seen at Trapp Lake.

Richie Lake is immediately south and upstream of Trapp Lake. Richie is the smallest, shallowest and most eutrophic of the chain and is heavily used by migratory waterfowl.

The field between Richie Lake and the highway is a good place to spot kestrel, meadowlark, savannah and vesper sparrows and long-billed curlew in spring.

Napier Lake, immediately south of Richie Lake, lies in a deeper more picturesque basin surrounded by steep grassland hills and rock outcrops. Several ponderosa pines around the edge of the lake provide excellent vantage points for bald eagle, red-tailed and rough-legged hawks and osprey. Golden eagles are often seen in the area.

**Km 33.2:** Check out the volcanic bedrock outcrop (Kamloops group) across the road from Napier Lake. Look for rock wren at the base of the slope.

**Km 35.4:** A series of small irrigation ponds on the east side of the road are fringed with cattails. They are one of the first to receive spring red-winged blackbirds and are often dense with migratory waterfowl waiting for larger bodies of water to open up.

The very picturesque Stump Lake Valley continues for the next 13 km. Grasslands along here are all privately held with the public restricted to the road corridor. Wildflowers that can be seen on the west side include fringe cup, rock cress, yellow bell, buttercup and shooting star. Hawks hunt the area from April to October. Rough-legged and marsh hawks coarse through in spring and fall.

Stump Lake contains at least 22 species of birds and at least a dozen species of fish. A band of bulrush surrounds about 75% of the lake.

The strip of land between the highway and the lake is rich in willow, rose, ceonothus, snowberry, rabbit-bush, and songbirds. In winter look for sharp-tailed grouse.

**Km 37.5:** In the field to the west are large volcanic rocks with cavernous hollows cased by weathering processes.

The large rock bluff on the east side of Stump Lake near the south end is known as Mineral Hill, the site of the Planet Mine where gold and copper were once mined. Seam and egg agates can be found in these outcrops. Creek beds on either side of the lake also contain pebbles of egg agate and jasper.

At the south end of Stump Lake muskrats can be seen. The field south of the lake is an excellent spot for yellow-bells in late March to early April.

**Km 49.9:** This is the beginning of a long sloping valley known as Guichon Flats. The top part of the flats is a mixture of open meadows and dense riparian habitat which are privately held. Bird life must be viewed from the road. Coyotes and a variety of raptors are frequent visitors to the flats. Balsamroot causes a spectacular yellow bloom in May. Old beach ridges from glacial Lake Merrit are prominent on the sides of the valley.

**Km 55.3:** The south end of the flats, just north of the ranch buildings, has been developed by Ducks Unlimited for breeding and migrating waterfowl. From ice-off to ice-on the area receives a tremendous array of waterbirds. In March and April look for canvasback, pintail, wigeon, both scaup species, redhead, Canada geese and tundra swan. On a good day several thousand water birds will rest on this wet area.

**Km 58.7:** A scan of the cottonwoods can often pick up a bald eagle in spring. Nicola Lake is a deep windy water body. In spring look for western and red-necked grebes, common loon, common merganser, an occasional white pelican or double crested cormorant. Flocks of gulls, scoters and snow geese pass on migration.

**Km 64.8:** The inlet of Nicola River to Nicola Lake opens up early in spring and is a good spot to watch for first arrivals of waterfowl. Look for kingfisher from spring to fall..

**Km 66.3:** ***Set your odometer at zero.***

**Km 0.0:** Take the Nicola Lake-Douglas Lake-Westwold public road which passes east through private ranch country. A drive along this road at any time of the year is well worth the time spent. In March and April, along the first 15 km, the fence posts are dotted with northern shrike and mountain bluebird. In the summer kestrel, vesper and savannah sparrows are very common. In spring, marsh hawk, golden eagle and rough-legged and red-tailed hawks hunt the open fields. Coyotes are common. Several osprey nests occur in cottonwoods along the river.

**Km 16.6:** Stop and scan the Douglas Lake outlet for waterbirds and shrubs for songbirds. Bald eagle and osprey are common along Douglas Lake in season. Eagles are best found at the east end of the lake in the cottonwood groves. Long-eared owls nest in the willows. Screech owls are rare in the cottonwoods along the river.

**Km 20.6:** East of the lake. A field to the south of the road often holds sandhill cranes in April and September, and Canada geese anytime from March to November.

**Km 21.7:** Entrance to the Douglas Lake Ranch home buildings.

**Km 21.9:** Just past the road into the home ranch, is Sanctuary Lake. Canada geese use it to breed, molt and stop over on migration. White pelicans occasionally come here, and common loons breed on the lake every year.

**Km 27.3:** The Upper Nicola River again crosses the road. Vaux swifts can be seen easily here. Herons nest downstream.

The next 6 km heading northeast from English Bridge at the Upper Nicola River, are the prime calving grounds for the Douglas Lake Ranch. During this spring period there is a tremendous raptor concentration as these birds feed on the afterbirth. Look for golden and bald eagles and rough-legged, red-tailed and marsh hawks. Coyotes are common too.

**Km 33.0:** On the south side of the road a large wild hay meadow is annually flooded by the Douglas Lake Ranch to produce more hay. The flooded field attracts huge numbers of waterfowl in April and May. Across from the meadow on the south hill look for several osprey nests.

Just east of the Big Meadow a hayfield on an Indian Reserve attracts long-billed curlew, mountain bluebird, northern shrike, sandhill crane, savannah and vesper sparrow. Badger dens occur in this area.

Chapperon Lake is immediately east and south of the Indian Reserve. The land is private but viewing from the road you can spot sandhill crane in April, ruddy duck and surf and white-winged scoter on migration. Bonaparte gulls sometimes number into the hundreds in spring migration.

**Birds:** common, black and common terns, bald eagles and osprey.

**Km 42.0:** Rush Lake lies on the north side of the road. For its size of 70 ha, it is one of the most heavily used lakes in the region by migrating waterfowl. Grain fields on the north side of this lake draw Canada geese and sandhill crane.

Salmon Lake is 1.6 km east of Rush lake. The west end of Salmon is a very productive marsh for water birds. During the several insect hatches over the summer this lake draws tremendous numbers of swallows, particularly barn. The field to the north, across the road from the lake, is a good spot for kestrel and rough-legged hawk from October to April. Swainson's and red-tailed are found here from April to October.

**Km 58.1:** Weyman Creek recreation site has dippers. Black bears are often seen along the road side in mid to late summer when the berries are ripe.

**Km 59.0:** A rock outcrop, grassy knoll and talus slope with pika and Columbian ground squirrel.
For the next 10 km the road continues to wind along the river with cottonwood, spruce and willow on the flood plain. Douglas-fir, juniper and some ponderosa pine are found in the higher areas.

***Return to Hwy 5A (set odometer to 0.0).***

**Km 68.8: (2.5)** **Birds:** tree, rough-winged and barn swallows, kingfisher, cedar waxwing, song sparrow, osprey, herring gull, red-winged, yellow-headed and Brewer's blackbirds, marsh wren and dusky flycatcher.

**Km 71.2: (4.9)** The historic Quilchena Hotel and general store. The fields just before the hotel are good locations to see Canada geese in spring and fall.

**Km 73.0: (6.7)** From here the road continues along the lake and provides a series of vantage points to observe waterfowl, eagles, ospreys, gulls and a possible river otter.
The adjacent hillside to the east was burned in the 1960's.

**Km 79.0: (12.7)** The lake outlet contains rich beds of bladderwort, northern watermilfol and pond weeds. Cliff swallows nest here in large numbers and house wrens are also present.
Chinook, coho and sockeye salmon migrate up the Nicola River in September and October and can be observed at the bridge.

**Km 83.3: (17.8)** Just past the bridge there is a turnoff to Monck Provincial Park (12 km). This is a popular camping and picnic area. The park receives an average of less than 26 cm of rain per year. Rubber boa may be observed on the park road in spring and summer.
**Birds**: nighthawk, osprey, common loon, white-breasted nuthatch, Lewis' woodpecker pygmy nuthatch and Townsend's solitaire.
**Mammals**: bats, northwestern chipmunk, yellow-bellied marmot, coyote, mule deer and porcupine.
**Reptiles**: alligator lizard, rubber boa snake and striped and dusky garter snakes. (None are poisonous.)
**Trees**: ponderosa pine, Douglas-fir and black cottonwood.
Hwy 5A continues south along the Nicola River to Merritt. The river provides habitat for numerous waterfowl including Canada goose, wood duck, hooded and common merganser, blue heron and a variety of songbirds.

*Ian Barnett*

# ADAMS RIVER AREA

The country east of Kamloops and immediately north of Hwy 1 offers a variety of opportunities to explore nature from near desert to lush vegetation along the shores of Adams and Shuswap Lakes. Every four years the sockeye salmon run up Adams River in hundreds of thousands (over 2 million in 1982). Over 190 species of flowers have been recorded including green-banded mariposa and chocolate lilies in July and August, and calypso and mountain lady slipper in May and June. Tracks of beaver and deer (mule and white-tail) are common along the shores and beaches. Birds to look for include golden eagle that nest at the head of Adams Lake and bald eagles and gulls foraging on the dead salmon every fall.

## Roderick Haig-Brown Conservation Area

This is one of the easiest locations in B.C. to see the spectacular salmon runs. Fish come in thousands every year but the huge numbers arrive every four. The males, brilliant red with pea green heads, have elongated hooked snouts, and are about 0.6 metre long. Females are also red but somewhat shorter. On the spawning grounds, the female selects a home territory in the gravel. The whole sequence can be watched while strolling along the bank. Salmon can be seen in numbers from about October 1 to 31, with the best times the second and third weeks of the month.

The best bird watching location in the whole area is at the mouth of the Adams River. The peak flowering time is April through June. Watch for poison ivy.

There is also a spectacular canyon in the Adams River, 1.6 km upstream from the Adams River Bridge.

**Birds:** bald eagle, osprey, loons, several species of gulls and ducks, American dipper, Canada geese, trumpeter and tundra swans, pileated woodpecker, northern flicker (both yellow and red shafted), mountain chichadee, rufous and black-chinned hummingbirds, several warblers and thrushes.
**Mammals**: whitetail and mule deer, beaver, mink, river otter, coyote, bob cat, moose, cougar, Woodland caribou, elk, grizzly and black bears.
**Plants:** lilies, orchids, yellow bells, grass widow and shooting star.

## Bear Creek Historic Log Flume Trail

There are several loops along this trail, allowing you to take a very short walk for a quick look at the flume or a 5-km hike around the longest loop. Remains of the flume can be seen along the way. Plants and mammals are the same as above. To reach this trail drive north-east from the Squilax overpass for 4 km to the Adams Lake turn off. Turn north or left for 5 km to a large parking area marked Historic Log Flume. Leave the car in the lot and proceed on the walk.

## Adams Lake

The 65-km-long lake is very deep in places and surrounded by mountains and tree-covered slopes.

Look for the golden eagle that nests at the head of the lake. Wild flowers abound from April to July.

**Further Information**
To reach the area, proceed east of Kamloops along Hwy 1 for about 50 km and watch for the Squilax overpass sign immediately east of Chase. Turn north off the highway and drive 6km along the Squila-Anglemont Rd to the various spots. • A provincial government boat launching site at the south end of the lake provides access to the water. • Refer to topographic map 82 L/NW Shuswap Lake, Province of B.C., Minister of the Environment. Check with the B.C. Outdoor Recreational Map for this area too. A marine map of Shuswap Lake may be obtained by writing the address below. • Contacts for further information may be made at Shuswap Lake Provincia l Park, Site 12, Camp 4, R.R. 1, Chase, B.C., V0E 1M0. Phone 955-2217.

*Elsie Nykyfork*

# SALMON ARM

Shuswap Lake with long arms and hidden bays, adjacent dry interior hills combined with wet cedar forests and alpine meadows offers opportunies for naturalists to watch one of only two colonies of western grebes in the province, numbers of Lazuli buntings, the northern most nesting colony of white-throated swifts, several species of butterflies and a variety of orchids including white reined and calypso.

Lying along Hwy 1 west of Revelstoke and east of Kamloops, Salmon Arm on the southwest corner of Shuswap Lake provides an ideal base to explore this area.

## Salmon Arm Lakeshore

Water levels in the lake fluctuate by as much as four metres from high years to low. The western grebes (one of only two breeding sites in B.C.) change their nesting location depending on the water level. During normal years the western grebes nest west of the wharf. They may be watched during courtship and nesting from mid-May to the latter part of June from the filled area by the sewer outfall. From March through to early May and August through mid-September the mud flats are alive with migrating shore birds.

**Birds:** bald eagles, ducks, sora and virginia rails, western, pied-billed, red-necked and horned grebes, yellow throat, yellow-headed black bird, marsh hawk, osprey and catbird.

The Salmon Arm area covers the shoreline from the wharf two km west to the delta of the Salmon River and one km east. The Wharf Rd lies in the middle west side of the community. Cross over the railway tracks on Wharf Rd, and proceed west from it along Beatty Avenue to Narcisse Street. Take Narcisse north about 100 m past the Salmon Arm sewage treatment plant to the entrance of the sewer outfall area. This filled area is raised about three metres above water level and provides a good spot to observe the marsh. Canoes may be launched back at the wharf.

## Gardom Lake

An island on this lake has nesting red-necked grebes that are readily seen and photographed. As many as six nests are visible from the less than a km long trail around the island. The conifers on the island and the deciduous forests on the main shore are full of small birds from May through June. In mid-June dragon and damselflies are numerous.

To locate the spot, proceed east from Salmon Arm Motor Hotel on Hwy 1 to the junction with 97B. Turn right and proceed on 97B to the Gardom Lake Rd. Turn right on Gardom Lake rd and follow it for 2.2 km to the "Park Rd" sign. Turn left and drive for a short distance to the park and parking area. There is a beautiful little picnic area by the lake. Access to the small island is by canoe or boat. A one-km foot path follows the perimeter of the island.

## Wallensteen and Kernaghan Lakes

These small lakes lie west of Salmon Arm in the Fly Hills area. From June to September this spot is worth the trip. The profusion and variety of vegetation supports a wide diversity of birds and insects.

**Birds:** hermit thrush and Lincoln's sparrows.
**Wildflowers**: kalmia and yellow violet, lupin, fireweed, pond lilies, white rein orchid and the tiny tway-blade.

To find these small lakes begin at the Salmon Arm Motor Motel in the centre of Salmon Arm. Take Hwy 1 west to the bridge over Salmon River. Leave the highway and proceed west along River Rd for 1.2 km to 13 Ave S.W. (Christenson Rd) and turn right; proceed 1.3 km to 5th Ave and turn left for 1.6 km to the sign "Forest Service Rd". Travel on weekends when there are no logging trucks. Proceed on this service road for 12 km to the Snowmobile Club hut. At the hut take the left turn and follow the road for about 5 km to Wallensteen Lake camping and picnic spot

## Sunnybrae Road to Herald Provincial Park

The 14-km road has convenient pulloffs at several spots. Begin at the Tappen turn off from Hwy 1, 11 km north and west of Salmon Arm. Drive east along the lake. On the north side of the road for the first 4 km the Sunnybrae Bluffs, reaching out of Douglas-fir forest, dominate the drive (see below). The next 4 km pass through parkland. Then it is 7 km around the base of Bastion Mountain to Herald Park. Bastion Mountain has one of the most northern breeding colonies of white-throated swifts in the world.

**Birds:** common loon, red-necked, western and horned grebes, common merganser, osprey, bald eagle, black and white-throated swifts, golden eagle, raven, red-tailed hawk, tundra swan, pygmy owl, goshawk, pine grosbeak, hoary redpoll and bohemian waxwing.

## Sunnybrae Bluffs Trail

The one-km trail to the top of the bluffs is a steep climb. It leads upward from lake shore through Douglas-fir forest to rocky bluffs and open ponderosa pine to the edge of an alpine meadow. Continuing around another one km you skirt a damp cedar forest then an alpine meadow, aspen parklands and back down through the Douglas-fir forest. On the way up keep left at the junction, as the trail loops back around to this spot on the way down.

As you hike the trail watch for butterflies, including yellow and black swallowtails, wood nymph, spring azure, mourning cloak, tortoise shell, fritillaries, orange tip and white admiral.

**Birds:** grebes, warblers, vireos, sparrows, kinglets, flycatchers, Hermit, Swainson's and varied thrushes, veery, osprey, western tanager, northern oriole, calliope and rufous hummingbirds, Townsend's solitaire, merlin, mountain bluebird, olive-sided flycatcher, red-tailed hawk, Lazuli bunting, varied thrush and flycatchers.

The trail begins and ends at Sunnybrae Provincial Recreation Area. Leave the TransCanada at Tapen and drive 4 km east to the parking lot. The land to the top of the bluff is owned by the provincial park. The top is under private ownership but no permit is required.

## Herald Provincial Park

At the park, opposite the parking lot, is a short trail to Margaret Falls. There you will find a deep ravine with vertical rock faces rising along either side of the trail. At the head of the ravine are the spectacular Margaret Falls. Look for dipper and winter wren at the falls.

On the way to the park below the big bluff, look for blazing star or mentzella, wooly thistle and white clematis. In ditches watch for giant helaborean. Beware of the poison ivy along the lake shore.

Wild flowers to be found in the park, from late April through August, include calypso and habenaria orchids, blue clematis, red columbine and aster.

The park lies east of Sunnybrae on the Canoe Point Rd.

## Little White Lake

This site lies north of Tappen and Shuswap Lake and is a fine place for marsh birds.

**Birds:** bittern, rails, marsh wren, grebes, ducks, warblers, vireos, tanagers, waxwings and flycatchers.

The small lake and marsh between this and the big lake is reached from Hwy 1 at the White Lake turn off 6 km north of Tappen. Drive north on this side road to where the road crosses the creek and marsh.

**Further Information**

A checklist of flowers, trees, shrubs and ferns of the Shuswap district by Mary-Lou Tapson-Jones was published and is available from The Shuswap Naturalist Club, Box 1076, Salmon Arm, B.C. V0E 2T0. • Maps include Recreation Map 7, Shuswap Lake region and topo map N.T.S 82L/NW, Shuswap Lake Region.

*Deane and Kay Munro, Mary-Lou Tapson-Jones,*

*Mary McGillivray, Frank and Doris Kime.*

# PRINCETON TO OSOYOOS – HWY 3

The road passes through desert benchland and dry open ponderosa woods to Douglas-fir forests. White-throated swift, Lewis' woodpecker Lazuli bunting, white-headed woodpecker, Brewer's and grasshopper sparrows and a wide selection of flowering plants including mariposa lily, scarlet gilia, sumac, antelope-bush, tumbling mustard, hairy golden aster and pink (showy) milkweed may be found in this cross section of habitats.

A traveller from the west will be struck with the change from sub-alpine forest to the open ponderosa pine parkland. Proceeding east the land takes on an even more open and arid appearance. Large expanses of grassland are dotted with sagebrush and juniper.

Along Hwy 3, Bromley Rock Provincial Park, about 22 km east of Princeton, and Stemwinder Provincial Park are two spots to get out and sample the surrounding natural history.

The "Old Hedley Rd" is an alternate and more pleasant route to take rather than the busy highway east from Princeton. To find this back road take the Merrit Hwy 5A out of Princeton, then turn off to the right to go past the large saw mill. There are two Forest Service picnic/campgrounds with access to the river along this route.

Driving along either route note the number of red-cedar trees along the banks of the Similkameen River. Some are quite old.

## Cathedral Provincial Park

The Ashnola River enters the Similkameen from the south about 9 km west of Keremeos. This is the gateway to the park and 33,272 ha of wilderness. Look for the turn off to the lower entrance, along Hwy 3 about 4 km west of Keremeos. Even if you don't want to go the full 26 km by gravel road to the entrance, a short drive up the Ashnola is well worth it.

The park habitat is typical of the transition between the rain forests on the west side of the Cascade Mountains and the arid Okanagan Valley. Plants and animals of the Cascades are different from the nearby Coast Ranges of the west and the Rockies to the east. The habitat within the park is mainly a drier alpine area. Mountain birds are numerous from May through July. Numerous lakes support rainbow and cutthroat trout.

**Birds**: Typical of the mountains.
**Mammals**: mule deer, California bighorn sheep and mountain goat, hoary marmot, golden mantled and Columbian ground squirrels.
**Trees:** Douglas-fir, aspen, cottonwood, lodgepole pine, Englemann spruce, subalpine (alpine) fir and alpine larch.
**Plants**: false heather and red alpine blueberry.

There are several hiking trails in this large wilderness area. Services are provided by a resort company. Primitive campgrounds occur along the spectacular Ashnola River on the way in, with the main park campground and trailer sites located at the end of the gravel road, just before the footbridge across the river.

A visitor may hike into the park or take the 4-wheel drive commercial trip into Cathedral Lodge, run by Cathedral Lakes Resort Company who have the exclusive transportation concession and access to the jeep trail via the private bridge across the Ashnola. Phone (604) 499-5848.

Hikers can usually arrange to have the Resort Company haul heavier gear to the lodge. Primitive camping is available across from the lodge at Quinisco Lake at a park campsite. Obtain map 92 H/SE Princeton sheet.

Back along Hwy 3, watch for the mountain goats on the cliffs north of the road in spring, early summer and fall. From the interpretive sign, about 12 km west of Keremeos, and on eastward look for birds of the ponderosa pine - bunchgrass habitat. These include calliope hummingbird, Lewis' woodpecker, western kingbird, Say's phoebe, black-billed magpie, pygmy nuthatch and Cassin's finch.

The village park at Keremeos lies at the south side of town and is often good for finding a variety of birds including nighthawk.

Continuing east on Hwy 3, west of the old Richter Ranch and Richter Pass, watch for the side road that leaves the highway and heads south to the border. Take this road to find sage thrasher, a possible long-billed curlew, and maybe a burrowing owl. Mariposa lilies are common on the hillsides to the west in late June and early July.

Richter Pass consists of open grasslands with scattered trees and several large ponds. White-headed woodpeckers and black-throated sparrows have been reported in the pass area. Check the water bodies for ducks and other water birds. The most easterly of these ponds, Spotted Lake, is noted for the large circular pans of salt crystals in summer. The lake has a high concentration of magnesium sulfate (Epsom Salts) and was a favorite bathing site for the native peoples in earlier times.

From the pass area a good road leads north from the highway to the top of Mt Kobau. This road was built by the federal government to service an observatory that was never completed. Take the drive to the top for the views and to explore around. About 5 km east of the pass, a road to the south leads uphill to the Blue and Kilpoola Lakes. This area is mainly privately held with some recent subdivisions, but provides good bird watching along the roadsides.

**Birds:** ruddy duck, red-tailed hawk, American coot, spotted sandpiper, common nighthawk, calliope hummingbird, Lewis' woodpecker, dusky flycatcher, Townsend's solitaire, warbling vireo, northern oriole and lark sparrow.

Stop at the lookout on the way down to Osoyoos for a good view of Canada's "Pocket Desert." Osoyoos Lake is only 277 m bove sea level, the lowest point in the southern interior, east of the Cascades. Note the narrow point of land extending almost across the southern tip of the lake, just north of the U.S. border. This is Haynes Point Provincial Park (see below). Road 22, another good nature route, is visible at the north end of the lake, north of the meadow where the river enters the lake. No 22 starts at Hwy 97 and proceeds east over a bridge towards the drybench lands on the east side of the valley. Look for burrowing owl along here. There is an interpretive sign here about these owls.

At the junction of Hwys 3 and 97, next to the Husky Station, check at the tourist information trailer for detailed directions to Haynes Pt Park, other camping sites and motel accommodation in the area. Campgrounds can be found at Haynes Pt and on the Indian Reserve just east of Osoyoos on the lake. To reach the reserve campground, drive east on Hwy 3 and turn left at the campground sign just a short distance past the windmill. As a guest of this campground you are allowed to hike around this unique desert area, go swimming or take part in guided horseback trail rides (no motor bikes or ATV's).

## Haynes Point Provincial Park

This 5 ha wave-formed sandspit, lined with stately cottonwoods, juts out into Osoyoos Lake.

The small marsh at the west end of the point and the sand bar on the eastern tip are the spots to watch for birds, particularly during spring and fall migration. A naturalist is available during July and August.

## Road 22 and the Ecological Reserve

Drive north on Hwy 97, for about 10 km from the Osoyoos business district. Take Rd 22 east over the bridge to the east side of the valley. Look for burrowing owls. Bobolink arrive about mid-May and leave about mid-August. This is one of the best sites to see these birds in the Okanagan.

The ranchers who hold the grazing lease do not object to naturalists on the land as long as they are on foot. A rather steep road may be followed by car along the vineyard fence to the north to reach the Ecological Reserve fence near the rock bluff. Park near the Ecological Reserve sign and walk along the track. Wildlife is often easier to see inside the reserve and nearer to the cliffs.

There are some interesting small mammals in the sandy desert including the great basin pocket mouse. These small animals are in the same family as the kangaroo rat of south western Saskatchewan and south eastern Alberta. These mice are rarely seen as they are nocturnal and remain underground to keep cool and conserve moisture during daylight hours.

A few rattlesnakes are found around rock slides and under bushes. Be particularly careful in April and May when they might be dozing on rocks after a cool night. Watch where you place your hands when climbing.

The main rock bluff is called "The Throne." From the bridge or the highway it appears to have a level spot about half way up that serves as a seat. Such ledges are good sites to find plants not grazed out in the over-pastured areas nearby. Antelope-bush occurs almost continuously throughout the Okanagan valley from Osoyoos to Kaleden and in patches as far north as Westbank. Cactus is very common so wear boots if hiking. Watch for poison ivy among the rocks. The local species of sumac is not poisonous and provides a great show of colour in the fall.

At the bridge across the river there are a variety of choices to explore further. Try the wet lands along the east dyke that heads south toward Osoyoos Lake. The bird watching is good here from April to October but the best time is mid-May to July. Most of the wetland area east of the channel is Crown land and shared as both a grazing lease and a Wildlife Management Area administered by B.C. Fish and Wildlife.

**Birds**: rock wren, canyon wren, lazuli bunting, lark sparrow, Lewis' woodpecker, American kestrel, nuthatch, eastern and western kingbirds and Clark's nutcracker black-crowned night-heron, sora, willow and dusky flycatchers, yellow-breasted chat, long-eared, short-eared and great horned owls.
**Mammals**: skunk, badger, coyote and mule deer.
**Plants**: sagebrush, rabbit-brush and antelope-bush or bitter-brush (known locally as greasewood), pallid evening primrose, blazing star, brittle prickly pear cactus and bitter root.

For further help contact the wardens of the Ecological Reserve at 495-6907.

While in the Osoyoos area, continue east from the town along Hwy 3 to explore several different habitats. The road rises through desert benchland, dry, open ponderosa pine woods to western larch and Douglas-fir woods, all within 20 km. Continue upward but stop at the lookout. Below you, this "almost desert" that you have come through is unique in Canada. Again you will see sage brush, antelope-bush and tumbling mustard.

On the way look for pink (showy) milkweed and sumac. In September and October the seeds of the sumac are a rich red while the leaves and stem turn a bright crimson. The hairy golden aster blooms from late spring through summer. Look for the scarlet gilia growing along the road between Rock Creek and the Okanagan Valley from May-July. This plant, common throughout the dry interior, attracts rufous and calliope hummingbirds.

On the way up there are several places to stop including the Highways Dept. rest stop, surrounded by pine woods and a nice aspen bluff. Ferruginous hawks have been found a few km east of here. Turkey vulture is uncommon but present. Deer are around in the early morning and Columbian ground squirrels are often found in sandy areas.

**Birds:** chukar, western meadowlark, magpie, violet-green or rough-winged swallow, white-throated swift, Lewis' woodpecker, red-naped sapsucker, dusky flycatchers, buteo and red-tailed hawks and lazuli buntings.

*Steve Cannings and Gail Moyle*

# VASEUX LAKE

This is the northern limit for any abundance of white-throated swift. White-headed woodpecker, pygmy nuthatch and canyon wren are present. This is a good place to see California bighorn sheep, and one of the few places in B.C. to find Nuttall's cottontail.

The lake and park lie 16 km south of Penticton airport.

Canada geese use the lake as a major breeding site. During peak migration in April and November large numbers of waterfowl pass through. White-throated swifts nest in the cliffs, one of the few accessible sites in Canada. Rare birds that come periodically include trumpeter swan and snow goose (fall and winter). Hummingbirds (4 species) are best observed along Irrigation Creek.

The most interesting mammal around Vaseux Lake is the California bighorn sheep. It is indigenous to this area and can be seen here more easily than anywhere else, especially in fall and winter. The rams spend the spring and summer further back in the hills east of the lake.

Look for Nuttall's cottontail on the slopes below the cliffs east of the highway.

There are more species of bats found here than anywhere in Canada. They can be seen at night and are easier to find using a sonic bat locater.

Large-mouth bass and western painted turtle inhabit the lake. Rattlesnakes live in the rock slides below the cliffs. They are not aggressive, but give them distance.

The vegetation in these areas is extremely varied. The best time to see it and avoid the heat is April, May and June.

**Birds:** golden and bald eagles, California quail, white-headed woodpecker, white-breasted, red-breasted and pygmy nuthatches, mountain chickadee, canyon wren, Say's phoebe, dusky flycatcher, turkey vulture, prairie falcon, screech and pygmy owls, poorwill (back roads at night), black-chinned hummingbird, white-headed woodpecker, and Clark's nutcracker.
**Mammals:** mule deer, white-tailed deer; and (rare) pocket mouse.
**Plants**: antelope-bush, rabbit-bush, balsamroot, phlox, bitterroot, prickly pear cactus, poison ivy, oregon grape, sumac, saskatoon, and mockorange.

**Further Information**

To reach the area from Osoyoos take Hwy 97 north via Oliver, or from Penticton take Hwy 97 south via Kaledan Junction. Greyhound bus services Oliver and Penticton. There is an airport at Penticton with regular airline service to Calgary and Vancouver. Three provincial parks provide camping nearby. The best is at Okanagan Falls, 8 km north of the lake.

*Steve Cannings*
*Gail Moyle and others*

# OSOYOOS NORTH TO PENTICTON —HWY 97

Take this route and side roads to find white-headed, Williamson's and Lewis' woodpeckers, mountain chickadee and a variety of wildflowers from drylands to alpine meadows.

The first stop, Rd 22, 10 km north of Osoyoos, is known as a good birding and botany area.

Deadman's Lake occurs about one km north of Rd 22. Several interesting birds have been sighted here including American bittern, and some shorebirds.

At Oliver, there are several side roads of interest to naturalists. Fairview Rd begins near the town centre, adjacent to the Village Hall, and heads west up the hill past the high school to the old mining town of Fairview. Continuing west up the mountain beyond Fairview takes you to the summit and then down along the Blind Creek Rd to Cawston on Hwy 3 west of Osoyoos. Before tackling this road, particularly the high part in the spring, check with the Department of Highways at 1-800-663-4997. Another route would be to turn north off the Fairview Rd at the sign "Fairview-White Lake Rd" which follows the Fur Brigade route used in the early 1800's. This route goes right through to Kaleden, missing Vaseux Lake and Okanagan Falls. There are two roads back to 97 on the way north. This makes a good loop trip north to Kaleden and then back down to Oliver (see below). A third alternative is the McKinney Rd that proceeds east from Oliver town centre past the hospital and on through the Inkaneep (or Inkameep) Indian Reserve toward the Mt Baldy Ski Area. The scarce white-headed woodpecker has been seen along this road, from about Km 8 and up as far as Baldy Creek. (Refer to map 82 E / SW, Penticton). The lower part of this road is paved.

Back on 97 proceed to Okanagan Falls. Take the logging road east up through the Weyerhaeuser sawmill just off Maple Street. Go on weekends as logging trucks use this route on weekdays. This road penetrates larch forest where you may sight Williamson's sapsucker.

The Allendale Lake Rd leads east up Shuttleworth Creek on the north bank. The road leaves Hwy 97 to go east from Okanagan Falls. Staff at the Esso Station near the OK Falls hotel will give directions. This is another route to search for white-headed woodpeckers about 3 km east of town.

On the west side of "OK Falls," stop at the Provincial Park just past the bridge over the Okanagan River. The rapid running water at this spot attracts numerous ducks, especially in fall, winter and spring. Look for bufflehead, Barrow's goldeneye, various mergansers, and California quail. The Falls were never very high but disappeared when the flood control dam was built. This is the easiest spot to find a dipper in fall and winter, and provides good birding opportunities all year. Search for Cassin's finch, northern oriole, catbird and various warblers. The least flycatcher is a rare summer visitor but has been found here At least 12 species of bats including the western big-eared, hoary, California and little brown bat are in and around this provincial campground. Activity is great in the evening as the numerous bats catch their meals of insects.

If you enjoy fairly steep winding roads, continue south past the park and up the hill to Green Lake, Mahoney Lake and out through Willowbrook to the White Lake-Fairview Rd. (see below).

## White Lake Loop

This road can be reached from further south on the Fairview road. Alternatively, from the Okanagan Falls bridge proceed north for about 4 km to the sign pointing west to the "White Lake Rd." Set your odometer to zero at this point. The first part of the route winds through dry ponderosa pine forest with spring sunflowers, brown-eyed susan, and stonecrop growing in exposed areas. On the way up, in small ponds look for cinnamon and blue-winged teal.

At Km 2.4 mountain bluebirds can often be seen; listen for chipping and vesper sparrows. Later in the summer the delicate asymmetrical seed head of the oyster plant can be spotted in the fields.

At Km 4.3, near the small marsh, listen for a common snipe winnowing. Watch for hummingbirds. Yellow-bellied marmots can be seen among the boulders near the road bank at St. Andrews golf course.

At Km 8.0 take the Oliver turnoff on the left to enter the White Lake basin, an expansive area of rolling sagebrush hills. Walking is easy through these hills but watch for cacti or the occasional rattlesnake.

White Lake is best known for the Dominion Astrophysical Observatory. Guided tours are usually available on Sunday afternoon during the summer. Naturalists know the area as one of the birding "hot spots" in the province. This wide sagebrush basin hosts rarities.

The lake is more like a prairie alkali slough. In some summers it dries up but many species of ducks, grebes and shorebirds use the lake. There are three smaller ponds up in the hills to the east of White Lake. A rough hiking trail leads to these water bodies. April and May are the best times to see flowers and birds. Badgers are sometimes spotted and coyotes howl at night. Look for yellow-bellied marmot around rocks near the road to the Radio Telescope on the way up to the bench.

In April watch for clusters of mountain bluebells and yellow bells. The elegant pinkish-white bitterroot and purple mountain phacelia are in bloom in early summer.

**Birds:** Brewer's and grasshopper sparrows, Say's phoebe, long-billed curlew, cliff swallow, gray partridge, chukar, sandhill crane (in migration), Wilson's phalarope, mountain bluebird, vesper sparrow, long-eared owl, western screech owl, red-naped sapsucker, hummingbirds, flycatchers and western meadowlark (one of the most common birds here).

Southward on the road, was once among the best places in B.C. to see poorwill at night, but now that the road is paved these birds are not as commonly seen.

At Km 15.3 take the paved road on the left (Green Lake Rd) and bear left at the fork at Km 16.7. Mahoney Lake at Km 19.6 and Green Lake at Km 21.4 often have breeding redhead and Barrow's goldeneye. Views are spectacular!

Return to Hwy 97 at the Okanagan Bridge.

Onward, north on Hwy 97 to Penticton where the beaches of Skaha and Okanagan Lakes beckon, look for several species of gulls, terns and shorebirds on migration together with numerous Canada geese and mallards in fall and winter.

The head of the Okanagan River, near the *S.S. Sicamous* paddlewheeler, and below the dam is worth a visit for the rose gardens and waterfowl.

Numerous side trips are available around Penticton. Drive through the orchards to Naramata and then on to the north a few km up the Chute Lake Rd into the ponderosa pine woods.

On the west side of Penticton, a longer trip leads up to Apex Mountain and the alpine meadows. Both the plants and birds to be found along here are worth the trip. A loop back can be made by returning via the Green Mt. Rd to Hwy 3A and Kaleden.

Visit the Chamber of Commerce office along Lakeshore Dr. on Okanagan Lake to obtain detailed directions for the above and other side trips. Checklists and information may be obtained from the South Okanagan Naturalists Club at P.O. Box 375, Penticton, B.C., V2A 6K6.

Further north on 97, at Summerland, there is a fine beach at Sunoka Provincial Park on Trout Creek Point. Stop here at the east end of the parking lot to find Lewis' woodpecker.

The Summerland Research Station has beautiful gardens, picnic facilities and numerous birds. Look for hummingbirds, Townsend's solitaire, mountain chickadee and others. There are viewpoints overlooking the Trout Creek canyon.

## Peach Orchard Road Loop

Turn east off 97 at the stop-light in Summerland, onto Peach Orchard Rd. Set the odometer to zero. At Km 1.4 look north beyond the Peach Orchard Campground to a spectacular stand of ponderosa pine growing on the dry hillside. This municipal campground is a delightful spot with tall pines and a small creek.

Across from the campground, turn north onto Ramsey Rd. This takes you through an "edge" area with arid hills to the north and lush shrubs along a small creek on the south. Such transitions are usually much more prolific in wildlife and plants than in either of the main areas. Look for warblers, cedar waxwing, red-shafted flicker, Steller's jay, pine siskin, California quail and eastern kingbird. Return back to Peach Orchard Rd.

Turn south at Km 1.9 and stop at the Municipal Beach to view common loon, mallard, ring-billed gull, Canada goose, killdeer, and a spotted sandpiper. The east-facing cliffs, west of the beach, are home to bank swallows. At the base of the cliffs, below the swallow nest holes, late blooming rabbit-bush thrives as well as wild rose and clematis.

At Km 2.9 the provincial government operates a fish hatchery for rainbow and brook trout. Belted kingfishers occur and western kingbirds perch on power lines across from the hatchery.

The next park, Okanagan Lake Provincial Park, lies a few km north and has a total of 156 campsites. A walk along the lakeshore to the north end is often good for birding. More than 10,000 trees have been planted, making this an oasis of green in an otherwise brown landscape.

Antler Beach Provincial Park is next along Hwy 97. There is a small picnic site with a good opportunity to look for a loon especially in fall and early winter. Close by, just before crossing the bridge at Deep Creek, turn left to reach Hardy Falls Park. This is a local park featuring a beautiful walk through deciduous woods along the creek to Hardy Falls — a distance of less than one km one way. Check for a dipper at the falls throughout the year. In late September or early October watch for a spawning run of Kokanee (land-locked salmon).

Continue on Hwy 97 to Kelowna. An interesting alternate route takes the "Westside Rd" which turns off west from 97 after passing Westbank and before 97 swings down to the Kelowna bridge. The Westside Rd is paved and follows the approximate route of the fur brigades of 150 years ago. The first stop along this route would be Bear Creek Provincial Park, a large park with 80 campsites on Okanagan Lake. The road continues north to rejoin Hwy 97 near the restored O'Keefe Ranch headquarters.

*Steve Cannings and Gail Moyle*

# NANCY GREEN PROVINCIAL PARK VIA GRAND FORKS TO ROCK CREEK —WY 6

Set in the dry interior, the Grand Forks area is locally known as the "Sunshine Valley" with an average of 1,500 hours of sunshine per year. Located along Hwy 3 east of the Okanagan, west of Trail and just north of the US border, the boundary country around Grand Forks is rich with bird life.

**Birds**: Say's phoebe, mountain and western bluebirds, meadowlark, six species of swallows, 8 species of warblers, western and eastern kingbirds, three kinds of blackbirds, cowbird, thrushes, western wood peewee, waterfowl, Lazuli bunting, western tanager and black-headed grosbeak.

## NANCY GREENE PROVINCIAL PARK

The park is located in the subalpine forest, along Hwy 3, 45 km east of Christina Lake. There is a nice walk around the lake. Scars from fire can be seen on the Douglas-fir. Old man's beard lichen hangs from older trees that survived the fire.

**Birds**: Steller's jay, gray jay, dark-eyed junco, pine siskin, raven, several species of woodpecker, blue grouse, mergansers, grebes, geese and gulls.
**Mammals**: deer, red squirrels, and Columbian ground squirrel.

## GILPIN WILDLIFE MANAGEMENT AREA

On Hwy 3 near the management area sign, 7 km before you get to Grand Forks, watch for a gate on the left or south side of the road. Follow the trail down to the wetlands formed by the oxbows of the Kettle River. Close the gates to keep live stock off the tracks. Scan across the marsh to the geese nesting mounds for waterfowl.

**Plants**: Ponderosa pine, toad flax (butter and eggs), white yarrow, arnica, balsamroot and ox-eye daisy.

## WARD'S LAKE WILDLIFE MANAGEMENT AREA

This is a good spot from breakup to freezeup for waterfowl. Watch for painted turtles and muskrats.

West of Grand Forks and the Grandby River turn north onto the North Fork Rd and drive 1.5 km to the Management Area. For the return continue north for about 15 km to the Hummingbird bridge, cross, and then turn south, back toward the city. Along the river watch for herons and ducks. In the fields check fence posts for bobolink.

## Rattlesnake Mountain

This spot is named after the rattlesnakes and bull snakes which still may be seen sunning themselves on the dry slopes and along the trail beyond the homes. The slag heaps of the former Grand Forks smelter provide upcurrents for soaring ravens and crows.

Begin at the eastern edge of Grand Forks where Hwy 3 crosses the Grandby River bridge. Do not cross the bridge, but turn north on the east side of the bridge and head upriver. On the north side of the city, turn east off the Grandby River Rd to Valley Heights subdivision on Rattlesnake Mountain. Return to the Granby Rd and then south to Hwy 3.

## Gilpin Range

The Gilpin Range, a semi-arid south facing slope north of Hwy 3, is one of the most important winter ranges for ungulates in the province. In March and April hundreds of deer are on these slopes feeding on new growth.

The area is west of Greenwood on Hwy 3 between Christina Lake and Grand Forks.

The boundary area has a wide variety of plants ranging from the wet communities along the rivers and lakes to the dry desert-like lands, forests, subalpine and alpine meadows.

Christina Lake lies in a geological fault left by the movement of plate tectonics in Tertiary time. The kettles from which Kettle River derives it's name may be seen along the trail that takes off south at the top of the hill 100 m west of the junction between Hwys 3 and 395. Take this trail down to the river to see these kettles caused by the scouring of the river through the bedrock. While at the intersection of these two highways check the rock cuts for numerous marine fossils exposed when Hwy 3 was blasted through. There are several other kettles in the area. Some were caused by glacial action such as the one at Ward's Lake.

**Birds**: Lewis' woodpecker, lazuli bunting, kestrel, meadowlark, vultures and bluebirds.
**Plants**: red with purple oat grass, while scarlet gilia, great mullein, mountain lady slipper, Richardson's penstemon, knapweed and Russian thistle.
**Mammals**: mule and white-tailed deer, Columbian ground squirrel, yellow-bellied marmot, coyote, black bear, elk, beaver, muskrat, mink, river otter, California bighorn sheep, moose, mountain goat, cougar, lynx, bobcat, short and long-tailed weasels, wolverine, marten, skunk and raccoon.

*Bob Sheppard and Michael Freisinger*
*(members of the Boundary Naturalists)*
*Gail Moyle*

# CASTLEGAR VIEWPOINT AND THE MEL DE ANNA TRAIL

To take the walk from the parking lot, hike up a short rocky crest to a gate, through a fence, along the mountain crest and on to wetlands. The 3.1 km trail follows the edge of the lake, or slough, where mallard, goldeneye, shoveler, bufflehead, hooded merganser and Canada geese are often encountered. Two small bird blinds provide good viewing opportunities. Watch for American redstarts, western tanager and possibly lazuli bunting along the trail.

This spot is located a short distance south of Castlegar on Hwy 3 towards Salmo.

## Castlegar to Nelson Road Log

**Km 0.0:** The flashing light above the highway at the junction of the airport road and Hwy 3A. Head north towards Nelson. Around the National Exhibition Centre, just past the Village, western meadowlarks and mountain bluebirds can be seen in the spring. Turn left at the flashing light to get to the N.E.C.

**Km 1.7:** Just before a bridge across the Kootenay River, a road to the left leads near the river where harlequin ducks may be seen. Follow trail along the river to old trees. Lewis' wood peckers nest here. This is a good birding area.

**Km 2.9:** On the hillside to the left, glacier lilies, lupins and syringa bloom in profusion in April and May.

**Km 5.9:** Just past this rock cut, on the river bank to the right, Canada geese and a variety of ducks are usually noted on or by the river.

**Km 18.5:** Junction of Hwys 6 and 3A. This is the starting point for the road log north to the Slocan Valley (see below).

**Km 19.7:** The two ponds on the left usually have a variety of birds particularly in the spring. Look for ring-necked duck, bufflehead, common and Barrow's goldeneye, a possible ruddy duck, killdeer, spotted sandpiper, red-winged blackbird, gray catbird, yellow and yellow-rumped warblers, and barn, tree and violet-green swallows.

**Km 21.0:** At the Kootenay Canal Generating Station sign, take the road to the right across the railroad tracks to the bridge across the Kootenay River. Stop at each end of the bridge and look for ducks, geese, dippers and river otters. The woods should have brown creeper, red-breasted nuthatch, red-naped sapsucker, common flicker and lazuli bunting.

Return to Hwy 3A or take this public road which is an alternate route to Nelson. The best birding lies between the bridge you have just crossed and the next one that crosses the Kootenay Canal. Watch for the parking spot before the bridge. An osprey nest can be seen on the power pole to the right along the canal. From the parking spot walk along the road to the left below the canal. Down the short incline is a beaver pond with both beavers and birds. A longer hike leads to the head of the canal, a large colony of cliff swallows and another osprey nest.
**Birds**: mountain bluebird, Say's phoebe, American goldfinch, eastern kingbird, western tanager, kestrel, pileated woodpecker, several duck species, osprey and the herons.

**Km 33.3:** Near the far end of this bridge and to the left, look for an osprey nest on an abandoned telegraph pole. The road to the right at the end of the bridge leads to the site of the old bridge - a good side trip to spot a variety of ducks including a pair of hooded mergansers. Follow the road beside the river for more birding including sora rail, blackbirds and a variety of warblers.

**Km 35.8**: Grohman Narrows park and picnic site on a small pond. A good spot for ducks and song birds.

### Nelson south to Salmo

**Km 0:** On Hwy 6 at the edge of Nelson. Watch for black bear, white-tailed and black-tailed deer along the road.

**Km 9.6:** White bog (or rein) orchids and Indian paintbrushes along the road and in the swampy area.

**Km 23.6:** Rest area with a waterfall. Birds: dipper, swallows and song sparrow.

**Km 26.6:** The road to the left leads to the old gold mining town of Ymir. The road rejoins Hwy 6 at Km 28.

**Km 33.3:** Hidden Creek campground with a possible nesting red-naped sapsucker. Check the end of the meadow below the campground for more birds.

**Km 39.1:** Village of Salmo. Eriw Lake lies 3.2 km west of Salmo. There are a variety of water and song birds at this site.

*Hazel and Jim Street*

## VALHALLA PROVINCIAL PARK

Mountain goats, grizzly bears, white-tailed ptarmigan and other high country animals are found in Valhalla. In some of the drainages, such as Nemo Creek, pristine forests are found with large Western red-cedars up to 800 years old and two or more metres in diametre. An ecological reserve at Evans Lake protects yellow cedar trees, usually seen only on the west coast, and recent finds of salal on the shores of Slocan Lake provide other examples of west coast vegetation found far into the Interior.

The park lies south and east of Hwy 1 and east of the Okanagan Valley.

### Slocan Lake Shore

Most of the lakeshore is a pristine forest varying from rugged bluffs to secluded sandy beaches. Indian rock paintings, thundering waterfalls, a diversity of wild mushrooms and small flocks of migrant waterfowl are some of the attractions. Ponderosa pine, Douglas-fir, juniper, yew mosses and ferns are some examples of the diversity of vegetation. In the spring the semi-open bluffs are beautifully festooned with early blooming flowers. An area of coastal salal is accessed by a trail at Cove Creek.

### Wee Sandy Creek

The creek rises on the slopes of Mt. Meers, deep in the park, where the waters flow into large Wee Sandy Lake. Cutthroat trout, the occasional heron, duck and shorebird are found at this high-country lake. A rudimentary foot path extends from Slocan Lake 13 km up the valley to within 3 km of Wee Sandy Lake. Watch for mountain goat on the rocky hillsides above the trail. Black bear and grizzly are present.

### Sharp Creek

Across from New Denver, take the Sharp or Glacier Creek trail to an alpine valley with tarns and meandering streams coming off the glacier. This steep but well-marked trail runs 9.5 km beside the creek, which has several tumbling waterfalls. Look for majestic old giant cedars, hanging valleys, and grand views of the lake and mountains to the east. The glacier is steep in places, but there is an easy route to the ridge. Look for snow fleas and ice worms.

### Nemo Creek

This drainage system has not been logged and is one of the last unexploited valleys in the Slocan Valley watershed. Extensive stands of climax forest exist, including cedar, hemlock, spruce, and subalpine fir, with trees having diameters of two or more metres a common occurrence. The main trail starts from the lakeshore about 50 m north of Nemo Creek, with a shorter loop trail along the creek. At the start, the creek enters the lake via some lovely waterfalls with a variety of mosses and ferns growing on the damp rocks. In the fall the forest here is a good spot to search for a variety of mushrooms including chanterelles. The "rock castles" of large erratic boulders covered in moss occur at 3.4 km on the main trail. The trail continues up the valley. The middle and upper valley are recognized for their vegetation diversity including the outstanding forests, coastal fern communities and goat and grizzly bear habitats. Once at the top, return on the same trail.

### Beatrice Creek

The creek rises in Avis and Demers Lakes adjacent to the main Divide of the Valhallas. The route of 12 km is along a good trail through rugged mountain country and wonderful old forests.

### Gwillim Creek

This creek is also known as Goat Creek because of the numbers of mountain goats frequently observed on the steep rocky walls of the canyon. Some of these high granitic domes resemble Yosemite Park in California. The trail begins just west of Slocan. Cross the Slocan River at the town of Slocan. Just after the crossing turn right onto the first road. The hiking trail starts about 30 m south of the new bridge at the outlet of Slocan Lake. The path follows through coniferous woods, past moss- and fern-covered bluffs. In spring these bluffs are covered with blooming avalanche lilies and several species of saxifrage. The trail eventually deteriorates in the headwater country. The mid-valley zone has a high grizzly hazard for hikers. The easiest access is from the southwest via a long drive up the Hoder Creek Logging Rd, where within an hour or two of hiking, a recently upgraded trail leads to the solitude and alpine basins of Drinnon Lakes and Gwillim Lakes. Look for swamp laurel and other interesting plants in the marly small bogs in the high country.

### Mulvey Creek

The area is frequented by black and grizzly bears particularly in mid-valley. Because of the high grizzly hazard this system should be avoided except for the Mulvey Basin, which is the most famous and spectacular of the alpine areas within the park. This area may be accessed via the Little Slocan road and then Bannockburn Rd which takes you up to a high ridge. Only experienced climbers should attempt the difficult traverse over the peaks into Mulvey. For further information contact B.C. Parks Kokanee Creek, R.R. 3, Nelson, B.C. V1L 5P6.

Within the park birds are quite numerous and varied. The Nemo and Mulvey valleys support rich mosaics of grizzly bear habitat and should be visited with extreme caution.

The alpine flora is most spectacular in the high meadows at various locations such as Mulvey Basin, Nemo Creek headwaters and Wee Sandy Lake region.

Metamorphic rocks make up most of the park. The southern portion is dominated by Reesor's Valhalla Gneiss Dome, centered on Mt Gladsheim, the highest peak in the Valhallas. Variations of these gneissic rocks associated with and frequently overlain by granitic rocks, are evident in the more spectacular summits of the area such as Mulvey and the Devil's group. Glaciers have also left their mark with huge rounded boulders in nearly every valley.

**Birds**: white-tailed ptarmigan, golden and bald eagle and osprey.
**Mammals**: mule and white-tailed deer, mountain goat, mountain caribou, moose, grizzly and black bear, lynx, cougar, coyote, wolverine, otter, mink, marten, fisher, ermine, marmots, red squirrels and pikas.
**Fish**: rainbow, cut-throat, eastern brook and Dolly Varden trout.
**Plants**: cliff romanzoffia and purple or opposite-leaved saxifrage.

**Further Information**
This is a "wild" park, and is not recommended for the novice. The park may be reached on Hwy 6 about 100 km north of Nelson and Castlegar. Highways 23 and 6 lead south for 140 km from Hwy 1 at Revelstoke. This route includes a free short ferry ride across the Upper Arrow Lake. A third route is to take Hwy 6 east for 230 km from Vernon. • To enter the park from the south drive north on Hwy 6, from it's junction with Hwy 3A, to just beyond the entrance to Passmore. At 15.3 km north from Hwy 3A take the road which crosses the Slocan River just beyond the gymkhana grounds. West of the bridge proceed north up the valley of the Little Slocan River to the two Little Slocan Lakes for good bird watching. This is uninhabited country where it is possible to sight bear and deer. • Continue on to the Hoder Creek Forest access road at Upper Little Slocan Lake and turn left (west). Signs indicate directions and distance to the park boundary via Drinnon Lake. This bush road takes you to the hiking trail that leads to the Gwillim Lakes. They lie in a valley in the heart of the high peaks at the southern end of the park. • Continuing north beyond the Hoder Creek turn off another 6-7 km, you come to another Forestry access road to the park via Bannock Burn Creek. This provides access to a route over the ridge in the vicinity of Gimli peak, Wolf Ears peak, Mount Dag and to Mulvey Lake. • Information on the park is available in New Denver from the Valhalla Trading Post at the foot of Main Street. Write the Parks and Recreation Division, Ministry of Lands, Parks and Housing, R.R. 3, Nelson B.C. V1L 5P6 or phone them at 604-825-4421. *Exploring the Southern Selkirks*, by J. Carter and D. Leighton and published by Douglas and McIntyre of Vancouver covers trails in the park and adjacent areas. • The Valhalla Wilderness Society is publishing an excellent map and pamphlet *A Trail Guide to the Valhalla Provincial Park and the proposed White Grizzly Wilderness area.* It will be available for $8.00 from them at Box 224, New Denver, B.C. V0G 1S0.

*Wayne McCrory and other members of the Valhalla Wilderness Society*
*Nancy Anderson.*

# KOKANEE GLACIER PROVINCIAL PARK

Kokanee has extremely rugged terrain, colourful alpine meadows in July, glaciers and high snowfall.

Sited in the southeast corner of the province, 29 km north of Nelson and west of Kootenay Lake, the park is easily accessible but out of the way for many travellers. There is only one road to the centre, with numerous trails through mountain meadows and forest and good bird watching.

The park is centered on a massive rock structure called the Nelson batholith of granite. This intrusion of rock was injected into the area in Jurassic time. As this mass cooled, solution-bearing minerals flowed out along some fissures to impregnate adjacent strata. Hence the occurrence of ore bodies in the area. Later uplift and erosion has exposed the batholith and nearby ore-bearing rocks.

At the Gibson Lake parking lot there is a 2-km trail around the lake. Watch for dippers, shorebirds, ducks, herons and geese. A resident beaver is active at the upper end.

There is a good chance of seeing mountain goats in Coffee, Woodburry and Silver Spray basins.

A 9.6 km trail begins at Gibson Lake and ascends to Kokanee Lake, Keen, Garland and Kaslo Lakes before reaching the Slocan Chief Cabin. A resident colony of hoary marmots inhabits the Kokanee Lake area.

**Birds**: mountain sparrows, warblers, Clark's nutcracker, gray jay, blue and Spruce grouse, ptarmigan and golden eagle.
**Mammals**: mountain goat, mule deer, black bear, grizzly, hoary marmot, pika, Columbian and golden mantled ground squirrel, marten and weasel.
**Fish**: rainbow, cutthroat and Dolly Varden.
**Plants**: hemlock, Western red-cedar, lodgepole pine, western and alpine larch, Engelmann spruce, alpine fir, dwarf huckleberry, white rhododendron and other alpine flowers.

**Further Information**
This is a wilderness park with no facilities. Travellers must be totally self-sufficient. There are two main access roads into the park with four secondary routes. These roads are not suitable for low-clearance vehicles, buses or those hauling trailers. The first route is to leave Nelson and drive northeast on Hwy 3A for 19.2 km, then 16 km of gravel road to Gibson Lake parking lot within the south end of the park. • The second route begins at Kaslo on Kootenay Lake and proceeds 5 km on Hwy 31A to the south turn off. It is a 24-km drive on gravel road to the Joker Mill site and parking lot in the centre of the park. A trail continues from the parking area south to Joker Lakes and another to Slocan Chief cabin. There are signs along the way and a park brochure to help you find your way. On the way to the Joker Mill site at 6.4 km a road and trail to the right or north lead off to the Flint Lakes and Mt Carlyle. Another road and trail take off to the left, at 15.5 km from the highway, up Sturgis Creek to the old Revenue mine. This overgrown trail goes through lovely patches of paintbrush and penstemon to an old mine site. • A third route into the park leaves Hwy 6, 39.0 Km north of the junction with Hwy 3A. This little-used route is along a logging road up Lemon Creek and ends at the park boundary at the beginning of the access hiking trail into the high country. The trail passes through a beautiful old forest past several old mines on the way to the gorgeous Saphire Lakes. Wild flowers abound. Watch for grizzlies! • A fourth route into the park takes off from Hwy 6 at Km 62.0 from the junction with Hwy 3A. This mining and logging road leads east to the boundary of the park via Enterprise Creek. A 10-km hiking trail is followed to Slocan Chief Cabin. The trail is a beautiful 4-hour walk through a deep, mature forest, climbing slowly to Tanal Lake. This is a fishing lake with a camping area. Again watch for bears. • At the park boundary on the way in, by the above mining/logging road, another trail leads to the north (left) up Paupo Creek in the northwest corner of the park. This 6-km route is quite steep in

places but quickly gets a hiker into an alpine basin with attractive plants and an old mine. • The Woodbury Creek access leaves Hwy 31 on the west side, 6.4 km north of Ainsworth Hot Springs. At 13.2 km in from the highway there is a parking lot from which two trails lead into the park: the one to Woodbury Basin and the new Cabin at 9 km; and the other to Clover Basin and the Silver Spray Cabin. Again watch for grizzlies from spring to early fall. There is no charge for use of these two huts. Also from the Woodbury Creek access you may sometimes be able to drive 15 km to a parking lot and the start of a 2-km trail to Sunset Lake. In season the flowers are great here. The lakes contain some ducks with nearby Clark's nutcracker and gray jay and even the odd eagle is sighted. • There are three cabin/shelters in the park and no developed campgrounds. Primus type stoves should be used at all times to prevent forest fires. The Slocan Chief Cabin, located south of the Joker parking lot and built in 1896, is available at $5 per person per night on a first-come basis. The cabin will accommodate 20 sleepers with 6 the maximum size for any group. Silver Spray Cabin in the Clover Basin near the northeast corner is available free for 4 campers. The newly erected Woodburry Hut at the head of Woodburry Basin sleeps 10 people comfortably and is free. • There are 1:50,000 and 1:125,000 maps available on portions of the park from Photo Sales, Ministry of Environment, Parliament Buildings, Victoria, B.C. V8V 1X5. or from the Government Agent, Courthouse, Nelson, B.C. V1L 4K3. A park brochure and a reprint of an excellent map made in 1925 showing all the trails is available free of charge from the Parks Division at R.R. 3, Nelson, B.C. V1L 5P6. You may phone them there at 604-825-4421.

*Nancy Anderson, John Carter*
*Kokanee Park brochure*

# SOUTH SLOCAN VIA NEW DENVER TO NAKUSP — HWY 6

Follow Slocan River and Lake to their head, then over a low pass to Nakusp on Upper Arrow Lake. Waterbirds abound and vegetation ranges from near rain forest to subalpine.

High rainfall produces heavy forest cover on fertile bottom lands and along side valleys at lower elevations. The upper valley end is wetter with the lower end drier and hotter. Here ponderosa pine predominates with interspersed lodgepole pine and larch, giving way to cedars and firs.

A few stands of old-growth forest remain, consisting of large Douglas-fir, western red-cedar and western hemlock, with some white and Englemann spruce on wetter sites, and larch and white pine on drier ones.

(Note: As roads are upgraded and curves taken out, the Km readings are changed.)

**Km 0.0:** Junction of Hwy 3A and Hwy 6.
**Km 0.8:** Small patches of blue brodiea (wild hyacinth) grow prolifically every spring near here.

**Km 1.4:** In the big pool along the Slocan River look for mallards, goldeneye and teal.

**Km 8.4**: This foot bridge across the Slocan River is a good spot to use to check for water birds.

**Km 18.7:** Good bird watching on the river and backwaters.

**Km 19.3:** The gneissic rocks exposed here are probably of early Mesozoic age and can be seen along with the Nelson granites on both sides of the river to a point north of Slocan. The Slocan River is a good place to pan for garnets. Ducks occur in the back waters.

**Km 22.7:** Lebahdo Flats with farmland and good birding spots.

**Km 34.5:** Perry Siding has good bird watching along the banks and in backwater sloughs.

**Km 39.0:** The road to the west (left) leads to hay fields and good spring birding and trails to Slocan River.

**Km 39.2**: Just across the Lemon Creek bridge a road leads to the west and down to the junction of the creek and Slocan River. Waterbirds come to the backed-up river in spring.

**Km 41.2:** The old Lemon Creek Lookout trail has flowers in spring.

**Km 44.4:** Gwillim (Goat) Creek canyon is across the river with cliffs used by numerous mountain goats.

**Km 46.5:** The Springer Creek Rd leaves the highway to the east, just before the sign "27 km to Silverton." This mining and logging road gives access to a large wilderness area. There are numerous roads through this country with branches and sub-branches leading into alpine and subalpine country with many flowers.

**Km 48.5**: Viewpoint just before the avalanche gates. In spring look for avalanche lilies, saxifrages, blue-eyed Mary, bluebells, lungwort, paintbrush, yellow monkey flowers, penstemon and forget-me-not. Search the trees on the other side of the lake for a small heronry.

**Km 50.7**: Turn-out at the garbage barrel. Indian pictographs occur on rocks along the shore but are best viewed by boat. The old road below provides a good walk along the steep rocky shoreline and through a tunnel. Interesting geology and spring botany.

**Km 53.2:** The end of the Cape Horn section of road. Look for ospreys flying along the lake shore and good geological cross-sections.

**Km 54.5:** In spring the woods around Memphis Creek have calypso orchids scattered throughout.

**Km 57.5 and 60.2** Old undrivable roads provide easy hiking access to the plants and animals of the deep woods. Ruffed grouse often can be seen eating clover.

**Km 61.0**: The Enterprise Creek crossing and logging roads give access to this beautiful brook trout stream and to trails in Kokanee Park.

**Km 66.2:** In the spring, white and green rein orchids grow along the ditches of this Red Mountain road, an alternate route to Silverton.

**Km 67.5:** View point. Flowers, squirrels and chipmunks are found here.

**Km 68.5:** Mule and white-tailed deer yard up here beside the highway in various spots in winter.

**Km 69.5:** Bannock Point Rd. A local forestry recreation park with a new trail through the woods to a beach. Spring orchids in the woods and waterfowl on water.

**Km 73.5:** Silverton. A mining road follows Silverton Creek east for about 18 kilometres, ending at the trailhead to Fishermaiden Lake. The drive and the half-hour hike into the subalpine lake are good for birds and flowers. The historic Wakefield trail leads to Silver Ridge and Forestry Look Out. Watch for signs. Check with local people for other roads and trails. Look for bear and cougar.

**Km 77.5**: New Denver. The lakeshore walk from Bigelow Bay at the north end of town, south to Carpenter Creek, or Bigelow Bay north for about 2 km to Molly Hughes Bay, provides chances for good bird watching and botanizing.
**Birds:** loons, harlequin ducks and dippers.
There are at least 12 Indian pictographs on the cliffs along the lakeshore with most rock paintings on the west shore and quite visible. Check with the Parks Branch for the map of their location.

**Km 79.6:** A road runs down to the left from this golf course road, leading to a lakeshore trail to New Denver. In spring many birds and flowers occur.

**Km 82.5:** Take a left turn to the lake and the mouth of Wilson Creek to see beaver. At Wilson Creek, look for harlequin ducks and calypso orchids in the woods in spring. Later find pyrola, coral root, Indian pipe, pipsissewa, queens cup, wild ginger, blue clematis, fringe cup and rattlesnake orchid.
A right turn off the highway leads to a 34-km drive on the east side of Wilson Creek. This gravel logging road parallels the highway to join the Nakusp highway just past Box Lake. Along the road there are three or more wilderness campsites, access roads and hiking trails the proposed Grizzly Wilderness Area and a spectacular waterfall on Wilson Creek. One road, the Hicks Creek Rd, begins about 2 km after leaving the highway and leads to the northeast (right) to about the 2000 m elevation. Here trails and secondary logging roads lead to the subalpine and alpine ridges in the White Grizzly proposal, which are noted for their spectacular flowers. This trail is also a good

place to see deer, bear and other mammals at higher elevations. Marked trails such as Dennis Creek and Dolly Varden are easy to find.

**Km 82.9:** A right turn takes you up the west side of Wilson Creek past a regional garbage dump. This is a good spot to find bears, ravens, crows and other scavengers. The area is good deer winter habitat. Watch for logging trucks. Mushrooms, including pine, are numerous in the fall.

**Km 83.2:** The protected harbour in front of Rosebery provides a haven for waterfowl. A lake net holds fingerlings of the large Kootenay Lake strain of rainbow trout (Gerrard rainbow) to be released when they are large enough to survive predation. From here the abandoned C.P.R. railroad can be walked to the lakeshore.

**Km 84.0:** At the hydro line and road, take a walk to the top of the bluffs where penstemons, saxifrages and other flowering plants live on the dry rocky cliffs. From this point onward is good white-tailed deer habitat.

**Km 91.1:** A Girl Guide Camp, and hiking access road to the head of the lake. An interesting creek with tuffa (porous lime secretions) and the Bonanza Creek marsh are found here. Canada geese, several species of ducks and shorebirds nest here. The marsh and road around it are on private property, but responsible naturalists are welcome. Kokanee salmon can be seen spawning in September.

**Birds**: tundra swan, great-blue heron, Virginia rail, Brewer's blackbird, willow and dusky flycatchers, several swallows, MacGillivrays's warbler, red-tailed hawks, American kestrels, wood ducks and woodpeckers including pileated.

**Km 92.1:** Another road to the marsh leads down across the old railroad grade. Just beyond the marsh road, the Shannon Creek forestry access road takes off to the west and continues for about 11 km to a divide with good botanizing. The road continues on into the Cariboo Creek drainage area, emptying into the Arrow Lakes at Burton.

A few km up the Shannon Creek road, the Wragge Creek logging road cuts off to the left. This rough logging road gives access to the high country and Wragge Lake, a subalpine lake with good birding and botanizing. Watch for the signs into the lake.

About 9 km up the Shannon Creek road a second road leads to the left and the Shannon Lake trail. Watch for the forestry signs. This is another lake area with good birding and botanizing. Watch for bears. Just beyond the Shannon Lake trail, a road cuts back to the right into a basin. It gives access to

Big Sister Mountain and basin which is excellent for mountain flowers and to rugged peak.

**Km 93.2:** Community of Hills.

**Km 100.7:** Check for wildlife along power lines. From here to Summit Lake, along the roadsides and the open hillsides, a herd of elk winters, staying until late spring.

**Km 102.8:** Trillium, rare to the area, grows under the birch trees to the right. A small herd of white-tailed deer often winters here. Skunk cabbage is common in wet places.

**Km 104.2:** At the old railroad grade and Bonanza Creek crossing look for yellow water lilies. A stroll along the railway right-of-way in either direction in the spring should locate rein orchids and false lily-of-the-valley in boggy areas. Both willow and alder flycatchers occur, together with waterbirds. The abandoned railroad grade is in the process of being designated as a recreation corridor.

**Km.104.7:** In the bluffs across Summit lake from the Kinsmen campsite, mountain goats may be seen in fall and spring. Birds: Canada geese, common loons, common merganser and common goldeneye.

**Km 109.1:** Summit Lake logging road connects with the logging roads from Arrow Lakes and the Shannon Creek road. Huckleberry picking and bird watching are good here. Franklin grouse are found. Other birds to look for include mountain and boreal chickadees, olive-sided flycatcher, gray and Steller's jays and Clark's nutcracker.

**Km 113.7:** To Box Lake campsite. Check for fall mushrooms.

**Km 114.7:** Take the five-minute hike along this trail to a small lake in the deep woods. Good birding and mushroom habitat.

**Km 116.3:** Box Lake has a small heronry near the south end. Look for osprey nests.

**Km 117.4:** The Wilson Lake road provides access to Wilson, Horseshoe and Kimball Lakes. Good birding and botanizing.

**Km 118.2:** The Brouse Loop road is a 4-km loop which has good birding in the forest, transition and open farm land. Look for meadowlark, bobolink, Brewer's blackbird, savannah sparrow and snipe in the fields; the coniferous woods have chestnut-backed and black-capped chickadees, Townsend's and yellow-rumped warblers, Swainson's thrush, western tanager, brown creeper, golden and ruby-crowned kinglets and pileated woodpecker; the deciduous bottomlands contain alder, willow and dusky flycatcher, western wood peewee, solitary, warbling and red-eyed vireos, MacGillivray's, Nashville and Wilson's warblers, Townsend's solitaire and several swallows. Near the creek search for great blue heron, kingfisher, dipper, common merganser, osprey, northern waterthrush and spotted sandpiper. Midway on the loop the Crescent Bay Rd takes off to the Burton-Fauquier highway and cuts off Naskusp. In the winter at least 7 species of owls have been recorded along this loop.

**Km 120.1:** The west end of Brouse Loop rd. The area is good birding country including the farming areas above Nakusp.

**Further Information**

The brochure *A trail guide to the Valhalla Provincial Park and proposed White Grizzly Wilderness Area* ($8.00) has an excellent map of the road north from Perry Siding (Km 34.5) to Summit Lake (Km 104.7). Available from the Valhalla Wilderness Society at Box 224, New Denver, B.C. V0G 1S0 • Maps of the area include 1:250,000 Nelson (82F and Lardeau (82K) Sheets; 1:126,720 Slocan (82F/NW), Naskusp (82K/SW). • More assistance at Parks and Outdoor Recreation Division, Ministry of Lands Parks and Housing, R.R. 3, Nelson, B.C. V1L 5P6. Telephone 825-4421. • Contact Kootenay-Boundary Visitors Association, Box 172, Nelson, B.C. V1L 5P9, telephone 354-4831.

*Nancy Anderson*

# ARROW LAKES —FAUQUIER TO GALENA BAY —HWYS 6 & 23

Take this drive to spot a variety of land and water birds throughout the year, together with deer and bear. The north end, Galena Bay is only 50 km south of Hwy 1 at Revelstoke.

Vegetation in the area ranges from deep lush forests at lake level to sparse alpine lands. It is a great place for mushroom hunting in the fall. The lowland forests are typical of the interior wet belt and somewhat similar to coastal forests, with Douglas-fir, Western red-cedar and western hemlock dominating.

Since the lands between high and low water levels are flooded part of every year and exposed for the rest of it, nothing grows. As a result it is a non-productive shoreline with little attraction for waterfowl and shore birds. However, numbers of killdeer take up territory on the barren beach every spring, and then have their eggs flooded when the water rises.

Coming in from the west on Hwy 6, take the free government ferry across to Fauquier. The map shows Needles on the west side, but this community was totally removed when the dam was built.

Upon leaving the ferry watch for the small rest area on the left. Just beyond the parking lot are two small sewage ponds which attract a variety of ducks during migration.

The power lines along the east side of the highway seem to attract ospreys.

There are about 15 nests between Fauquier and Nakusp, but only a few are used each year. These large fish-eating birds are common along the route from May to September.

Between Fauquier and Burton, rough-winged swallows nest in the rock cuts on the east side of the road.

The highway runs on a causeway in the approach to Burton. Three creeks, Caribou, Snow and Burton, enter the lake via the bridge at this site. This fresh water estuary draws numerous migratory birds to feed in spring and fall. Hundreds of gulls come in September to eat the kokanee that have died after spawning in the creeks. Backcountry roads lead up all three creeks to access the sub-alpine areas.

Rising above the lake on the east side just north of the ferry crossing, Saddle Back Mountain has a forestry lookout at North Peak. A marked route leads to the top.

At Nakusp check out the hot springs Up Kuskanax Creek and the Cedar Grove, which has the largest cedars in the interior of B.C.

Leaving Nakusp, the road crosses the Kuskanax Creek. In late August and September this creek is a major kokanee spawning site. Gulls and a few ospreys are usually present.

There is a government fish hatchery for kokanee on Hill Creek just north of Galena Bay.

On this last stretch there are a number of hot springs, all with crude "homemade" pools. Directions to Halcyon, St Leon and Halfway hot springs are available from the Nakusp information office. None of these sites are signed or developed.

**Birds**: Barrow's and common goldeneye, all three teal species, wood duck, ring-necked duck, American coot and red-winged blackbirds, California, herring, mew and ringed-bill gulls, ducks, common loon, several grebes, tundra swan, white pelican, rock wrens, black-throated sparrow, American dippers and belted kingfishers; in winter pileated woodpecker, Steller's jay, common raven, finches, pine siskin, red crossbill, pine grossbeak, rosy finch, common redpol and Cassin's finch.
**Mammals**: Otters, mountain goats, black bear and mule deer.
**Wildflowers**: saxifrage, draba, sedum, moss campion, partridge foot, sandwort, fleabane, lousewort, dryas, alum root, harebell and dwarf willow.

**Further Information**
Information may be obtained from the tourist information booth in the village office at Naskusp. The forestry office will provide a map showing some trails and forestry campsites.

*Gary Davidson and Nancy Anderson*

# NEW DENVER TO KASLO —HWY 31A AND NORTH HWY 31

This highway provides access to forests, subalpine and alpine country in some of the more remote areas of the province.

From the end of June to late July alpine flowers are the highlight of the trip. There is relatively easy access into several alpine basins. Later in the season, along margins of creeks, search for yellow willow herb. In fall a variety of mushrooms occur along roadsides and in the forests.

The area is underlain by the Slocan Group of Upper Triassic age rock. These rocks have undergone considerable deformation and contain a mix of rock and structure types together with some mineralization. Silver, lead and zinc are still being prospected.

**Birds**; dippers, harlequin ducks, common mergansers, bald eagles, blue and ruffed grouse, white-tailed ptarmigan and hummingbirds.
**Mammals**: mountain goat, caribou, cougar, lynx, black and grizzly bears, marten, mink, pika, wolverine (winter), pika, hoary marmot (spring to fall), pack rats, beaver and muskrat.
**Wildflowers**: yellow avalanche lilies, white anemones, montane buttercups, globe flowers, marsh marigolds, lupine, Indian paintbrush, arnica, asters, daisies, delphinium, yellow and red columbines, forget-me-not, red and yellow mimulus, mountain rhododendrons, and opposite-leaved purple saxifrage.

(Note: As roads are upgraded and curves taken out, Km readings are changed, often considerably.)

**Km 0.0:** Junction of Hwy 6 and the west end of 31A in New Denver. As you begin the climb east watch for the cemetery on the east side of the road for lazuli bunting and calliope hummingbirds. New Denver is one of the few spots where the calliope is as common as the rufous.

**Km 1.4:** The old railway bed provides easy walking in either direction for spring and fall birding or winter cross-country skiing. It is 6 km north on this bed to Roseberry; to the east it is 3 km through the spectacular Denver Canyon of Carpenter Creek (watch for dippers) to the Alamo crossing.

**Km 3.1:** A very steep trail to the north climbs about 1.5 km to the old Capella mine and then on to Carpenter (Goat) Mountain. In winter from here for the next 5 km along the highway mule deer occur. A cougar is sometimes spotted.

**Km 8.0:** A mining road, usually in fair condition, up the creek the proposed White Grizzly Wilderness Area goes 12 km up the mountain to the old MacAlister mine. Past the mine, the road is sometimes in poor shape and is best done on foot. This road, and an adjacent old trail, continue up to London Ridge, a wonderful place to see alpine flowers from mid-June to mid-

July. In summer deer and caribou can be seen on the talus slopes. The area to the north of the ridge is grizzly bear habitat. Both sides of the ridge have good crops of huckleberries, but watch for bears. If you have two vehicles, or an extra driver, you can establish a pick up at Km 17 (see below) and hike over London Ridge on the trails to Vimy and Panama mines, then on down to Hwy 31A at Fish Lake.

**Km 8.4:** The historic ghost town of Sandon turn off. Just past the junction (watch for the Alamo Trail sign), an old mining road cuts back to the right and parallels the river back downstream on the old C.P.R. railway grade to the west, for about 3 km to Alamo. This is an enjoyable spring walk for ferns, other plants and birds.

West from Sandon, a road (watch for signs to Idaho Lookout) leads about 11 km to Wild Goose basin and on to Silver Ridge and a parking area where the historic Wakefield trail from Silverton Creek joins the ridge. From this trailhead it is an easy 1.5 km hike to Idaho Lookout for spectacular alpine flora, birds and mammals. Flying ants and ladybeetles often collect on the peak. The ridge provides easy access for star gazing on a clear night. East of Sandon a road leads up to Cody, another ghost town, and other old mining roads which lead to abandoned workings and provide access to sub-alpine.

**Km 8.7:** Access via a short steep pack trail to Payne Bluffs on the K&S Railway Historic Trail. A sign at the trailhead tells the story of this trail. A good trail guide and historic poster ($18) are available from the Valhalla Wilderness Society on the history of the railway from here to Sandon. This is a good trail to take in late spring or early summer. Morels occur in spring. This trail to and from Sandon plus a return hike on the Alamo trail makes a good day's outing.

**Km 10.2:** Access to the old C.P.R. railway grade. Alpine fir, white and Engelmann spruce are starting to appear.

**Km 13.0:** Rambler Mine Rd to several basins, mining prospects and old mines. This side road improves after the initial fording of the creek and provides access to this mountain area for hiking.

**Km 13.6:** Beaver pond with dams and houses. Check for dippers in the creek where it is fast moving just below the sign.

**Km 15.2:** Zincton town site (Lucky Jim Mine) with an old beaver dam, pond and house.

**Km 15.5** Bear Lake. The avalanche slopes around this and the adjacent Fish Lake are good for grizzlies, avalanche lilies, waterleaf or ball head, globeflower, meadow rue (carried down by the avalanches) as well as twisted stalk, delphinium, larkspur and Solomon's seal mixed in with willows and alder. In the spring take the old road around the lake to the left to see mountain rainbow and cutthroat trout spawning in a small stream that comes in from the avalanche gully.

**Km 17.0:** Road on the north side goes to London Ridge (see Km 8.0).

**Km 19.9:** Ghost town of Whitewater (Retallack). To the right of the avalanche gates the Jackson Basin Rd goes about 12 km into the alpine country. Just across the bridge off the highway and up the creek is a superb heritage grove. Just beyond the old mine and mill (at about 8 km) the road forks. The left route runs into several basins and ends with two small tarns set among lovely alpine vegetation. The right forks goes for 3 or 4 km to the ridge and another old mine site. As you approach the ridge from Jackson Basin you have easy access to Reco, Texas and Paddy's Peaks, and a spectacular walk along the ridge. Alpine flowers are especially beautiful from late June to mid-August. Look for the craters dug by grizzlies as they search for food, including ground squirrels, glacier lily corms and abandoned prospectors' diggings.

**Km 20.1:** As you drive through Retallack, the road leaving the highway on the north side provides access to other alpine areas. Drive beyond the buildings to reach the mining road up the mountain. At the first cross roads take the right fork to Whitewater Mountain and Glacier part of the proposed White Grizzly Wilderness area. This is about a 7 km trip. The road ends at a well-marked trailhead. The newly rerouted trail runs parallel to a long alpine valley with a wide variety of plants and birds Watch for grizzlies in the meadow across the valley. There is a campsite about 5 km up: please use the food/garbage hanging tree as grizzlies range all over. The glacier is not large with an easy climb past several blue-green tarns to the pass. Green serpentine rock is abundant at the foot of the glacier. This is prime grizzly country so be careful.

The Lyle glacier and Mount Brennan fork is well posted. Turn off to the right and follow the signs up the Lyle Creek road to alpine country. At the trailhead at the Highland Surprise mine follow the good trail to a hanging valley with a lovely blue-green lake with the color caused by glacier silt. From here, above timberline, it is relatively easy to follow the old mine trail to the left that disappears in the rocks here and there. This is mountain goat country with alpine flower rock gardens.

**Birds:** white-crowned, fox and golden-crowned sparrows, hermit thrush, water pipit, white-tailed ptarmigan, rosy finch and ruby-crowned kinglet.

**Km 21.8:** Lyle Creek crossing. Watch for dippers. As you descend from the summit and proceed to Kaslo, the road follows the small but fast-moving Kaslo River. Watch for merganser, dipper and harlequin duck.

**Km 29.2:** The left road goes about 16 km to Blue Ridge where there is good subalpine natural history and several old mines. One of the mines has deposits of linerite, a rare mineral of sulfate, copper and lead origin. It is a beautiful blue-green color. Ask at the mining recorder's office in Kaslo for directions to locate this mineral. The right (southwest) road goes up Twelve Mile Creek to the old Utica mine. Watch the bridge, it is in need of repair. From the mine a good hike may be taken to the ridge connecting Paddy Peak with Texas and Reco Peaks to the north and thence back down to Retallack.

**Km 33.8:** In early spring, just as the snow leaves, check this avalanche slope for grizzlies.

**Km 42.0:** On the left are "soda" springs containing carbon dioxide and iron. The red iron stain will help locate the deposit.

**Km 45.4:** Upper Kaslo on Kootenay Lake.

**Further Information**

There are few tourist facilities, or other amenities along Hwy 31A which is a wonderful back country road with side roads into the mountains and superb wilderness areas to see a great variety of plants and animals close at hand. Make sure your gas tank is topped up and you have spare emergency supplies along, particularly in the off season when traffic is sparse. • *A Trail Guide to the Valhalla Provincial Park and proposed White Grizzly Wilderness Area* ($8.00). An excellent map of the road east from New Denver to the Twelve Mile Creek side road at Km 29.2. • *The K&S Railway Historic Trail from Sandon to Three Forks, B.C.*, ($3.00). The accompanying map is quite useful for locating natural history spots around the trail. • Both brochures are available from the Valhalla Wilderness Society at Box 224, New Denver, B.C. V0G 1S0 or by stopping at the Valhalla Trading Post in New Denver. • Maps of the area include 1:250,000 Nelson (82F and Lardeau (82K) Sheets; 1:126,720 Slocan (82F/NW), Kaslo 82F/NE), Trail (82F SW), Lardeau (82K/SE). • For further information contact Parks and Outdoor Recreation Division, Ministry of Lands Parks and Housing, R.R. 3, Nelson, B.C. V1L 5P6. Phone them at 604- 825-4421. • For additional information contact the Kootenay-Boundary Visitors Association, Box 172, Nelson, B.C. V1L 5P9, telephone 604-354-4831.

*Nancy Anderson*

# CRESTON VALLEY WILDLIFE AREA

The Creston Valley, sandwiched between the Selkirk and Purcell Mountain ranges, is one of the most important migratory bird staging sites in interior B.C. Wetlands are uncommon in the interior of B.C., making this a very special place. In March 1500-3000 tundra swans use the area. Thousands of ducks come to the wetlands in April. Songbird migration reaches a peak in May. In summer Creston boasts the densest population of ospreys, black terns, grebes (western, red-necked, eared, pied-billed and horned) and probably sora rails, in B.C.

The valley supports the second largest breeding colony of western grebe, the second largest great blue heron colony east of the coast range, the densest nesting population of osprey (over 40 breeding pair from the US border north to Kootenay Lake), the only nesting site of American avocet and Forster's tern and the largest breeding colony of black tern. A check list of birds is available at the Centre.

The Kootenays are one of the best places in Canada to find bats, and many may be found at Creston.

The western painted turtles at the Interpretive Centre may be as old as 60-80 years.

As in all wetlands, insects and related creatures are numerous. Twenty-one groups and species of butterflies are found within 25 km of Creston. They are described in a draft booklet available from the Interpretive Centre. A short study of dragonflies and damsel flies found 28 species of dragonflies in the area.

**Birds**: blue grouse, dusky flycatcher, Steller's jay, Clark's nutcracker, mountain chickadee, varied thrush, Townsend's solitaire, Townsend's warbler, Cassin's finch, pygmy nuthatch, golden-crowned sparrow, white-tailed ptarmigan, wood duck, cinnamon teal, Barrow's golden eye, harlequin duck, dipper, long-billed curlew, mountain bluebirds, wild turkeys, calliope, rufous and black-chinned hummingbirds, lazuli bunting, and white-fronted geese.
**Mammals**: river otter, wood rat, hoary marmot, fringed bat, elk, caribou, beaver, muskrat, mink, black bear and moose.
**Reptiles and amphibians**: rubber boa, western skink, northern alligator lizard and western painted turtles.
**Fish**: bullheads (catfish), pumpkinseed, perch, bass, Cutthroat trout, mountain whitefish, rainbow and Dolly Varden trout.
**Plants**: Western red-cedar, western hemlock, grand fir, birch, cottonwood, red-osier dogwood, ocean spray, ninebark, thimbleberry, Douglas-fir, ponderosa pine and western larch.

### Summit Creek Campground

Dewdney Trail starts from the suspension bridge in the picnic area. Garter snakes can sometimes be seen warming themselves on exposed rocks. Just beyond the bridge, the purple flowers of *clarkia* have chocolate tips. On warmer evenings little brown bats fly through the campground on their hunt for insects.

**Plants**: Western red-cedar, birch, western hemlock, black cottonwoods, thimbleberries, ponderosa pine, western larch and Douglas-fir.
**Birds**: winter wren, red-eyed vireo, pileated woodpecker, Swainson's thrush, American dipper, Western tanager, rufous-sided towhee, mourning dove, blue grouse, red crossbill and evening grosbeak.

### Char Basin

Fire burned several hectares of forest in 1970. Scan the standing snags for woodpeckers including pileated, hairy, downy and flicker. Lupines bloom along the highway.

### Stagleap Provincial Park

The park is noted for the herd of Selkirk Mountain caribou that sometimes frequents the Kootenay Pass. These mammals summer in the high alpine country. The population is down to 20-25 animals today.

Spire-shaped alpine fir dominates the forest canopy surrounding Bridal Lake. These stands are mature, over 120 years old, and support lichen. Non-forested areas dominate the ridges and high elevation slopes.

**Mammals**: Selkirk Mountain caribou, mink, marten, moose, snowshoe hare, Columbian grounds squirrels, pika and hoary marmot.
**Birds**: pine siskin, gray jay, Clark's nutcracker and evening grosbeak.
**Plants**: alpine fir, rhododendron, mountain ash, huckleberry, false box, menziesia, heather, penstemon and bear grass.

**Further Information**

Creston lies on Hwy 3, east of Trail. The Interpretive Centre is located on the west side of the Wildlife Management Area, 12 km west and south of Creston. • For further information write The Centre at Box 1849, Creston, B.C. V0B 1G0; or The Management Authority Box 640, Creston, B.C. V0B 1G0 or phone 428-3259 or 428-3260. They have several publications which are provided upon request.

*Gail Moyle, R. Butler, J. Sirois*

# KOOTENAY LAKE AREA

A flock of tundra swans winter at the south end of Kootenay Lake, common loons and western grebes are fairly abundant in the warmer months, and river otters have been spotted at the mouth of Kootenay River and along the east shore of the lake. This is the home of the largest rainbow trout in the world. They spawn near Gerrard every spring. Thousands of kokanee salmon spawn in most tributaries in late August.

Lying in the south east part of the province, Kootenay Lake occupies about half the length of a steep-sided valley extending 225 km in a northerly direction from the U.S. border to west of Golden. The valley is bounded on the east by the Purcell Mountains and on the west by the Selkirk Mountains. The west arm of the lake discharges to the south into the Kootenay River which flows into the Columbia.

The Creston Wildlife Interpretive Centre lies on the Kootenay flats at the south end of the lake.

The head of Kootenay Lake is a waterfowl sanctuary. Deer and elk winter here and the land has been set aside for their protection. North from Howser, on the

Lardeau River Rd, the John Fenser Memorial Trail park area has been set aside as a place to walk in a virgin forest among giant red-cedars, western hemlock and Douglas-firs.

North of Meadow Creek, just past the end of the pavement on the west (left) side, an old marble quarry is almost hidden by trees. Bats can be seen in the main cave.

### Kokanee Creek Provincial Park

Kokanee Creek has carved a rock-girdled path through the western part of the park and ends in a sandy delta on Kootenay Lake. The park entrance is on Hwy 3A, 19 km east of Nelson.

**Birds**: Ospreys, great blue heron and rufous hummingbird.
**Mammals**: squirrels, chipmunks, beaver and muskrat.
**Fish**: kokanee salmon.
**Plants**: Birch, Douglas-fir, lodgepole pine, cottonwood, Western red-cedar, western hemlock, larch and alder.

### Ainsworth Hot Springs

The springs were known to the Indians as both beneficial to them and to the deer that came to bathe their injuries, well before white men arrived.

The walls of horseshoe cave have been growing coats of different minerals for centuries. Hot, steamy, odorless mineralized water falls from the roof into a pool of waist-deep water, providing a natural steam bath.

The caves are open all year and are located about 48 km north of Nelson or 14.4 km north of the ferry landing at Balfour.

### Cody Caves Provincial Park

The Cody Caves contain an impressive display of calcite formations which have been growing at the rate of about one cm per century. They began forming about 170 million years ago when these mountains were thrust upward.

In late fall daddy long-legs or harvestmen literally cover the left wall of the Entrance Chamber as they move in to winter here.

The soft, white, putty-like material seen on the walls is moonmilk. Consisting of calcite and bacteria it is made of distinct crystals suspended in microbiological material. Some forms of this deposit contain a component with antibiotic properties.

Many forms of calcite deposition may be seen as you walk the corridors. Look for stalagmites, soda straws, bacon strips, helictites and draperies.

The Park may be reached along a narrow rough forest road that leaves the west side of Hwy 31 at a small gravel pit about 3 km north of Ainsworth. This 15-km road is passable from June through October for most high clearance vehicles. Proceed along the main road, ignoring all secondary routes, to the main junction at about 9 km from Hwy 31. Take the south fork which is marked with a directional arrow. Cross Cedar and Krao Creeks to the parking lot where there is a sign for the caves, which lie about 0.8 km up the trail.

### Meadow Creek Spawning Channel

This man-made spawning channel is easily the world's largest and the only freshwater kokanee spawning channel. Every fall between 1.3 and 1.5 million kokanee make the run up Meadow Creek. The following spring about 10 million fry leave the creek and channel for the lake.

The Meadow Creek channel is located 43.2 km north of Kaslo.

### Gerrard - Home of the Giant Rainbow Trout

Gerrard is noted as the spring spawning grounds of the large Kootenay Lake rainbow trout. To see them go to the head of the Lardeau River at the mouth of Trout Lake near the old community of Gerrard. Here about 1000 of the largest rainbow trout in the world come to deposit and fertilize their eggs.

Migration from the lake begins late in April and is over by late May. Most of the fish move by night and reach Gerrard in about 13 days. Peak numbers occur in early May.

Several thousand whitefish come down from Trout Lake to scavenge trout eggs.

The trout grow quickly to weigh as much as 16 kilograms, with an average weight of 8 kilos. The largest Gerrard rainbow ever captured weighed 23.6 kilos (52 lbs). The largest sport caught fish from Kootenay Lake was 16.1 kilos (35.5 lbs), taken in 1975.

The trout return to spawn from 4-6 years of age. The rigors of spawning usually exhaust them, and most die. One remarkable fish lived 14 years and spawned at least 3 times. Most are fortunate to live beyond 3-4 years, with some surviving 8-10 years.

Gerrard is located along Hwy 31, about 80 km north of Kaslo.

**Further Information**

There are several campgrounds in the area and suitable accommodation along the southern part of the lake. North of Kaslo there are few services. • Tourist facilities are few in the Lardeau area. Trout Lake, at the north end of the lake close to Ferguson, has the old historic Windsor hotel, a laundromat, motel and gas station.

*Gail Moyle*

# ALBERTA BORDER — CRANBROOK —KIMBERLEY — CRESTON —HWY 3

Beginning in the heart of the Rocky Mountains, and extending west through the Rocky Mountain Trench and the Purcell Mountains to the Wildlife Sanctuary at Creston, this route covers everything from dry prairie habitat to Douglas-fir forests and marshy wetlands. This East Kootenay country has the largest number and highest density of big game species in the province and probably one of the highest in North America. Hwy 3, from Pincher Creek, Alberta in the east, is the southernmost route across British Columbia.

## Sparwood

There is a large herd of non-migratory elk here.

## Morrissey Provincial Park

Take a short walk to the Elk River to scan for common mergansers. There are large cottonwoods along the bank.

The Rocky Mountain trench lies to the west. This valley is part of the very long depression that extends from Montana to the Yukon and divides the Rocky Mountains from a series of ranges to the west. At this south end, the Purcell Mountains form the western flank of this elongated basin.

The floor of the trench has a dry climate. It is also a major flyway for migrating birds.

## East of Elko

The highway passes through dry ponderosa pine country which gives way to Douglas-fir, larch and trembling aspen. Elk come down from the high country to this dry area to winter.

## Kikomun Creek Provincial Park

The park entrance lies south of the junction of Hwys 93 and 3. Take 93 south and then the road west into the park. The park lies in the bottom of the Rocky Mountain Trench. The dry climate, combined with the logging and fires in the early 1900s, have enabled bunch grass and ponderosa pine to dominate what otherwise would be a Douglas-fir landscape. Open grasslands provide excellent winter range for deer and elk. Numerous birds are present including waterfowl, woodpeckers, owls and songbirds. Besides the big game look for skunks and beaver. Flowers include brown-eyed susan, yellow pond lily, elegant mariposa lily, balsamroot and gromwell.

Hidden Lake, within the park, is a kettle type of lake occupying a depression left by a glacier. The trail around it takes about 25 minutes. Look for common goldeneye, western painted turtle and large dragonflies.

### West side of Rocky Mountain Trench

Just north of Wardner, there is a good view of the trench to the east. This landform extends north almost to the Northwest Territories border.

### East of Cranbrook, Hwy 3

The lands along the highway lie in a dry rain shadow forest that covers much of the drier valley bottoms. Ponderosa pine and bunchgrass dominate at lower elevations, giving way to Douglas-fir and western larch further north and higher up the slope. Common birds include flicker, western tanager, mountain chickadee and mountain bluebird. Elk and deer are commonly seen. Badger signs, consisting of large half-moon-shaped holes, can be spotted. Voles and pocket gophers inhabit the grasslands.

### Cranbrook

The city was built on open prairie in the arid Rocky Mountain Trench. Kinsmen Park, in the south central part of the city, with its large cottonwoods, willows, alders and red ozier dogwood is the best place to enjoy the creek, trees and birds.

The Tourist Information Office is at the west city limits. It abuts on Elizabeth Lake and marsh wildlife area. Ringed by bullrushes, the marsh is a prime nesting area for waterfowl and other water birds. Moose and elk can be seen there in season.

### Kimberley

The Kimberley Ski Resort, partly within the city, has a chairlift to the Alpine Slide which provides opportunities to reach otherwise hard to get at alpine.

### 15 km south of Cranbrook

Continuing south on Hwys 3 and 93, watch for small open pockets of water on the west side of the road about 15 km south of Cranbrook. Look for a roadside turnout. Watercress is the green plant that forms the dense mat with white flowers in bloom most of the summer. Its leaves are edible and often used in salads. The water stays open most of the winter. Goldeneye are often here in the cold weather. Look for great horned owls that perch around the open water.

**Birds**: Macgillivray's warbler and rough-winged swallows.
**Plants**: red osier dogwood, oyster plant (*Tragopogon dubius*) and a blue borage.

### Moyie Lake Provincial Park

At 19 km south of Cranbrook, this park is lightly forested with Douglas-fir, larch, lodgepole pine and white spruce in the drier areas, while black cottonwood, trembling aspen, mountain alder and willow grow near the water. The marshy area, where the Moyie River enters the lake on the eastern side of the park is a good site to observe waterfowl and various species of warblers. Kokanee (landlocked salmon) gather at the mouths of the Moyie River and Peavine Creek each fall, to spawn in the gravel beds.

"Kettle Pond Trail" is a pleasant self-guiding walk.

Moose feed on marsh vegetation in the shallow waters beside the road south of the park.

**Birds**: pileated woodpecker, red-naped sapsucker, brown creeper, solitary vireo, Swainson's thrush, pine siskin and red crossbill.
**Mammals**: varying hare, Columbian ground squirrel, white-tailed and mule deer.
**Fish**: brook and rainbow trout, Dolly Varden and kokanee.
**Plants**: self heal, kinnickinnick, shrubby cinquefoil, pyrola and harebell, Oregon grape and the blue fruit.

## Moyie River Rest Stop

About half way between Moyie Park and Yahk watch for the rest stop on the right (north) side of the road. Turn left and look for tiger lilies and lupines in early summer.

Western larch and Englemann spruce are the dominant trees. The long stringy lichen hanging from the trees is old man's or young man's beard (depending on the species). Check on the ground for twinflower.

**Birds**: golden-crowned kinglet, Swainson's thrush, yellow-rumped warbler, northern water thrush and song sparrow.

## Kidd Creek Rest Stop

This stop lies west of Yahk a little over a third of the way to Creston, on the north (right) side of the road. A short walk to the creek may reveal a dipper.

**Birds**: pine siskins, red-breasted nuthatches, juncos and yellow-rumped warblers
**Plants**: lodgepole pine, grand fir, Douglas-fir, Western red-cedar, and black cottonwood, alumroot, Oregon grape, mountain spirea and snowberry.

## Skimmerhorns

The vertical range of mountains on the south side of the highway, just east of Creston, is the Skimmerhorns. A herd of mountain goat was transplanted there several years ago and are holding their own. Mountain bluebirds nest in the boxes placed in the alpine meadows by the Canadian Wildlife Service. Watch for harlequin ducks in the Goat River along the road for the last 19 km into Creston.

## Creston Valley Flats

The 12-km road from Creston to the Wildlife Centre is through lands reminiscent of the open prairie. Just after turning south off Hwy 3 to the Wildlife Centre, watch for Columbian ground squirrel beside the road. Be on the alert for female western painted turtles crossing the road in May and June. They lay their eggs in soft, sandy, gravelly soil along the road bed. Coyotes can often be heard "singing" in the area east of the Centre.

**Birds**: bank and cliff swallows, kestrels, cinnamon teal, blue-winged teal, bobolink, western meadowlark, Brewer's, red-winged and yellow-headed blackbirds, magpies, turkey vultures, red-tailed hawks, ring-necked pheasants, mourning doves, white-fronted and Canada geese, rough-legged hawks and flickers.

*Gail Moyle, Leo S. Gasner, Q.C.*

# GOLDEN—RADIUM—CRANBROOK —HWYS 95 AND 93 / 95

This route follows the Rocky Mountain Trench, one of the most distinctive features in North America. The Trench is a 3-16 km wide, flat-bottomed valley that extends in a remarkably straight line from Montana to northern B.C. Its 1600-km length makes the trench the longest of its kind in the world.

The Columbia River wetlands are unique. No other interior wetlands in the province is of comparable size or supports such diversity of wildlife.

The marsh between Canal Flats and Golden is important for duck staging, and Canada goose production and is a major breeding area for bald eagle, osprey and great blue heron. Two hundred and seventy-three species of birds have been recorded for the area.

The valley also supports large populations of elk and deer and is an important wintering area for Rocky Mountain bighorn sheep.

The trench is also one of the best places in the province to observe the northern leopard frog.

The mountain ranges greatly modify the climate. The Purcells to the west create a rain shadow by acting as a barrier against Pacific air. The Rocky Mountains on the east tend to protect the valley from cold Arctic air that covers the prairies in winter. This drought condition produces the dry interior Douglas-fir bioclimatic zone.

The Purcell Mountains feature high peaks with extensive areas of alpine glaciers and icefields. They are composed of metamorphic and sedimentary rocks consisting of argillite, quartzite, granite, slates and dolomites. The Rocky Mountains on the other hand, are composed of sedimentary rocks consisting mainly of limestones and shales.

**Mammals**: Rocky Mountain bighorn sheep, moose, mountain goat, black bear, grizzly, cougar, bobcat, muskrat, beaver, coyotes, Columbian ground squirrels, red squirrels, mink, weasels and American badger.
**Amphibians**: western toad, spotted frog, northern leopard frog, wood frog and northern long-toed salamander.
**Reptiles**: common gartersnake (red-sided subspecies), western terrestrial gartersnake (wandering subspecies), painted turtle and rubber boa.
**Plants**: Douglas-fir, ponderosa pine, white spruce, western larch, Western red-cedar, Englemann spruce, subalpine fir, lodgepole pine, aspen, black cottonwood, birch, willow and alder.

The route is divided into two sections: Radium north to Golden, Hwy 93 and Radium south to Cranbrook, Hwy 93/95.

## Radium Hot Springs south to Cranbrook, Hwy 93/95

**Km 0.0:** Junction of Hwy 93 and 95 at Radium Hot Springs.

**Km 1.7 to 2.6:** Columbia River view point. The Columbia River channel bounded by natural silt levees which support a dense growth of black cottonwoods, willows, alder and red ozier dogwood. Belted king fisher, northern rough-winged swallow and bank swallow nest in the clay banks above the railway track. Deer and elk, particularly in fall, winter and early spring occur along here. A walk on the terraces will locate early spring flowers such as townsendia, double bladder pod, scentless chamomile, crocus, shooting star and narrow leafed parsley. This area is also the most northern limit in the trench for ponderosa pine. **Birds**: yellow warbler, gray catbird, cedar waxwing, black-capped chickadee, mallard, wood duck, common goldeneye, bufflehead, pied-billed grebe, red-necked grebe.

**Km 4.8:** Dry Gulch Provincial Park. Flying squirrels have been reported here.

**Km 7.6:** This open Douglas-fir forest on the lower slopes of the Rockies is a major winter range for Rocky Mountain bighorn sheep and mule deer. It is also an important area for predators such as coyotes and cougars. Limber pine and prickly pear cactus may be found on these south-facing slopes.

**Km 13.5:** Invermere Junction. Access to Windermere Lake and James Chabot Provincial Park.

**Km 19.5:** Bald eagles congregate at the mouth of Windermere Creek to feed on spawning kokanee in September and October.

**Km 21.0:** Between Radium and Fairmont the islands of grasslands surrounded by open stands of Douglas-fir and Rocky Mountain juniper, and aspen provide attractive habitats for red-tailed hawk, northern flicker, ruby-crowned kinglet, yellow-rumped warbler, mountain bluebird and vesper sparrow.

**Km 31.0:** Dutch Creek Burn is viewed to the west. Fire occurred in 1972 and again in 1985 and has provided homes for many cavity nesting birds including Lewis' woodpecker.

**Km 41.0 to 42.8:** The hoodoos created by rainwash are landform features of semi arid country. Watch for nesting white-throated swift on these clay cliffs. These birds are one of the fastest flying birds in North America.

**Km 41.5:** West Side Rd junction- Access to Dutch Creek Burn and Invermere.

**Km 41.7:** In fall Rocky Mountain whitefish spawn in Dutch Creek.

**Km 47.0:** Viewpoint. The Columbia Lake was formed by the damming effect of the Dutch Creek alluvial fan. Gravel deposits from this fan create a major spawning area for burbot. The lake is also the headwaters of the Columbia River.

**Km 50.0**: Numerous osprey nests, like the one on this Douglas-fir snag, can be seen while driving the highway. This bird is the most abundant raptor found in the Columbia valley. The river and lakes in the valley, with their ample supply of northern squawfish and suckers, support one of the largest populations of ospreys in B.C.

**Km 55.0:** The open southwest-facing slopes on the east side of Columbia Lake are important winter range for Rocky Mountain bighorn sheep.

**Km 56.0:** A mixed forest of Douglas-fir and ponderosa pine.

**Km 58.5**: At Thunderhill Provincial Park, saw-whet owls may be heard calling in the spring and early summer.

**Km 58.8**: Beds of bulrush occupy much of the shallow area of the south west corner of Columbia Lake and extend narrowly along the south shore where they mingle with stands of cattails and cane grass. During March and April this area is a favourite resting spot for migrating tundra swan, Canada goose, green-winged teal, mallard, northern pintail, American wigeon and common goldeneye.

**Km 59.9:** A prominent beaver lodge is visible in the pond west of the highway.

**Km 61.2:** Canal Flats.

**Km 68.0:** In this mixed forest check for western larch and ponderosa pine with an understory of buffaloberry, birch-leaved spirea, common juniper and pine grass.

**Km 68.2:** Access to Whiteswan Lake Provincial Park and Lussier Hot Springs.

**Km 79.0:** Island Pond. Spring and summer birds seen here include Barrow's goldeneye, redhead and canvasback.

**Km 89.0:** These cleared hills are one of the most likely areas for seeing elk near the highway, particularly in spring. The ribbon on fence wires is to make them more visible to wildlife when crossing.

**Km 95.0 to Km 100.5:** the Skookumchuck prairie provides field habitat for long-billed curlew.

**Km 95.8:** The fenced enclosure on the west side of the road helps biologists determine long-term effect on these meadows by the grazing of livestock and wildlife.

**Km 105.0:** Junction to Kimberly.

**Km 107.8:** Entrance to Wasa Provincial Park. Boating, swimming, fishing, camping, a self-guiding nature trail and summer naturalist are available here.

**Km 111.5 Km 115.5:** Wasa Slough Wildlife Sanctuary, is an excellent spot for birding. Look for great blue heron, turkey vulture, common nighthawk Canada geese, sandpipers, kingbirds, osprey, bald

eagle, northern flicker, and hairy and downy woodpeckers. Also a good spot for painted turtles.
The sloughs (mostly in plain view from the highway) are home to many insects such as dragon- and damsel- flies and water tiger beetles.

**Km 116.5:** The open hillsides produce ponderosa pine, common rabbit-bush, antelope-bush, bluebunch wheatgrass, rough fescue and prairie koeleria. This is home to the pygmy nuthatch.

**Km 120.4:** Bummers Flats are best seen from the crest of a long hill. The round potholes were created by Ducks Unlimited for water-fowl. Canada geese, mallards and other surface feeders, ruddy duck, black tern, Wilson's phalarope are quite common. Sometimes turkey vultures can be spotted overhead. Prairie crocus is the first spring flower followed by phlox. Milkweed blooms in mid-June and is fairly common along the steep banks below the road. Sandhill cranes have been reported migrating through this area.

**Km 129.3:** Junction to Kootenay Trout Hatchery. Aquarium exhibits tell the story from fish egg to fingerling.

**Km 129.6:** Fort Steele Provincial Historic Park.

**Km 130.5:** The black cottonwoods growing on the river bank are able to survive periodic flooding, unlike most trees.

**Km 133.5:** Ponderosa pine forest.

**Km 139.5:** Mixed forest of western larch and Douglas-fir. The increase in available moisture has allowed western larch to grow. The understory is dominated by buffaloberry, saskatoon and pine grass.

**Km 142.8:** Junction of Hwys 95 and 93/3.

## Radium Hot Springs north to Golden — Hwy 95

**Km 0.0:** Junction of Hwy 95 and 93/95. Entrance to and information on Kootenay National Park.

**Km 0.2:** Before the Columbia River dams were built in the U.S.A., salmon travelled 2000 km from the Pacific Ocean to spawn here in Sinclair Creek. Bighorn sheep are occasionally seen on the open slopes. These mammals are confined to the Rocky Mountain side of the Trench. They summer on the alpine meadows and winter on the lower open slopes facing southwest.

**Km 12.2:** In spring western kingbirds sometimes nest on poles with electrical transformers.

**Km 14.0:** This Christmas tree farm land provides excellent range for ungulates, particularly in March and April.

**Km 16.5:** The roadside provides habitat for common mullen, sow thistle, common fireweed and ox-eyed daisy.

**Km 21.0:** "Cauliflower Tree" is the name given to this strange-looking Douglas-fir. It is suggested that a virus caused this unusual growth.

**Km 28.8:** Turn off to the west at the Brisco Rd, to the Columbia Wildlife Area (1.8 km) and to Bugaboo Glacier Provincial Park (45 km). At the Columbia Wildlife Area, the Canadian Wildlife Service installed a number of nest boxes for wood duck and common goldeneye.

Just west of the Columbia River at this site, two sloughs lie on either side of the road. The south one is dominated by yellow pond lily and is not particularly attractive to waterfowl. The northern slough supports a heavy growth of bulrush and is a favoured spring resting site and breeding area for waterfowl. Look for wood duck, green-winged teal, cinnamon teal, northern pintail, northern shoveler, American wigeon, redhead, bufflehead and ruddy duck. Black tern, marsh wren, red-winged blackbird all nest here. Wilson's phalarope may be seen during migration.

**Km 29.0** A control burn was held on the island to clear dead wood and improve willow growth for ungulates. Burning around marshes prior to the nesting season also builds up the roots of plants eaten by ducks and geese.

**Km 31.7:** A great blue heron rookery can be seen in the tall poplars between the two river channels. The 300 pairs of herons nesting in the Upper Columbia river valley constitute the second largest concentration in Western Canada.

**Km 33.6: to Km 40.0:** White spruce becomes more conspicuous as you proceed north. This tree shares the uplands with Douglas-fir. The bottom lands are covered with aspen, cottonwood, birch and willow.

**Km 40.0:** Spillimachine and access to the Columbia River.

**Km 42.1:** An osprey nest.

**Km 50.0:** The open waters of Hot Creek attract wildlife here in the dead of winter.

**Km 54.1:** Muskrats nest in the centre of the large mounds of vegetation and mud found in the marshes.

**Km 61.7 to 103:** Note scattered nesting platforms for Canada geese. Duck production in the Columbia Valley is low because of late spring and early summer flooding. Their nests are inundated by high water before the broods hatch. However, geese nests are quite successful because most broods are hatched from nests set on high ground, on muskrat houses, in osprey nests or on man made nesting platforms throughout the marshes. These platforms also isolate nests from most predators.

**Km 72.0:** Western red-cedar grows here in wet areas.

**Km 73.0:** Many American kestrels are seen perched on the power lines during spring and fall migration.

**Km 90.0:** Moose are found along the tributaries draining the eastern slopes of the Purcell Mountains where there are avalanche paths or old burn areas.

**Km 103 to 104:** Reflection Pond, a great spot for waterfowl.

*Larry Halverson*

# GLACIER AND MT REVELSTOKE NATIONAL PARKS, ADJACENT SITES: GOLDEN — SICAMOUS

Hwy 1 cuts across the Rocky Mountains to Golden, through the Rocky Mountain Trench and on through the Columbia Mountains as it makes its way west to Sicamous and on to the coast. The road provides excellent opportunities to explore Yoho to the east and then Glacier and Mount Revelstoke National Parks.

Because Glacier and Mount Revelstoke National Parks are snow bound for seven months of the year, large mammals are sparse, although small mammals and birds are abundant. Animal travel is difficult, and food is covered deep under the snow.

Birds include the high mountain species. Mount Revelstoke Summit Rd has four species of chickadees, including black-capped, mountain, boreal and chestnut-backed, warblers, blue grouse, black and Vaux's swifts, three-toed woodpecker and olive-sided flycatcher. At the top watch for fox sparrow. A hike through the summit meadows can also yield hawk-owl, golden eagle, white-tailed ptarmigan, water pipit and rosy finch.

East of Revelstoke Park junction, the stopping sites should produce several birds. In winter watch for influxes of finches obtaining salt and grit along the road.

No bighorn sheep occur in the Columbia Mountains. Herds of mountain goat can be seen on some cliff faces in the Selkirks. Grizzly and black bears flourish throughout the Columbia Mountains.

These mountains are one of the last major refuges left in the world for the grizzly. Travel by people is limited and hence there are few bear/people contacts.

Many small mammals are common, including pika and hoary marmot.

The Columbia Mountains have three broad types of vegetation. The cedar/ hemlock forest along most of the roadway changes to subalpine at about 1600 metres. At about the 2000-metre mark the trees disappear giving way to alpine tundra.

Timberline occurs at about 2000 metres, lower in the Columbia Mountains than in the Rockies because of the longer time the snow covers the ground.

In the main valleys, luxuriant forests, similar to those seen on the west coast, consist of western red-cedar, western hemlock, mountain hemlock, Engelmann spruce, Douglas-fir, and western white pine. Understory in these forests is littered with deadfall, and there is little light penetration. Few plants grow except moss and Devil's club with a resultant shortage of feed for browsing animals. Wet areas have luxuriant skunk cabbage.

The rocks that make up these mountains were laid down in a shallow sea that covered Western North America about 1.6 billion years ago. They were put under tremendous pressure and temperature. The layers were compressed, melted and transformed into hard metamorphic rock. At the same time other foreign rock was injected into these layers. About 160 million years ago, these now-metamorphised rocks were folded and raised to the surface to form the mountains of gneiss, schist, quartzite and slate with intrusions of granite. From then until now they have been worn down by the glaciers and eroded by rain, snow, running water, wind and cold.

There are few fossils in the Columbians as most of the rocks were laid down before recognizable life forms appeared.

In contrast, the Rocky Mountains to the east were also laid down as sediments in seas but much later when life forms were well developed. The Rockies were uplifted near the end of Cretaceous time about 65 to 75 million years ago. The Columbias are made up of limestone, mudstones, sandstones and shales with many fossils throughout, some world famous.

**Birds:** most species of B.C. ducks, gees, and swans (Revelstoke), northern Goshawk, merlin; barred, great horned, saw-whet, pygmy and hawk owls; black and Vaux swift; veery; Swaison's, hermit, and varied thrush; black-capped, chest-nut-backed, mountain, and boreal chickadees; black-headed gros beak, red and white-winged crossbill; pine siskin.
**Mammals:** grizzly and black bear, caribou, white-tailed and mule deer, elk, moose, mountain goat, pine marten, wolverine, many species of bats, American pika, Columbian ground squirrel, hoary marmot.

**Km 0.0:** Golden. The Rockies to the east are characterized by wide sweeping valleys and distinct peaks. The valley floors hold drier forests of Englemann spruce, subalpine fir and lodgepole pine with a bountiful and varied animal life. The wide-bot-tomed valley called the Rocky Mountain Trench separates the Rockies on the east from the Columbias to the west. The trench ranges from 3-16 km in width and stretches in a remarkable straight line from Montana almost into the Yukon, for 1600 km, making it the longest trough in the world.

**Km 16.0:** To the west is a view of the eastern side of the Columbias.

**Km 20.8:** An excellent view down the Rocky Mountain Trench. This valley lies in the rain shadow of the Columbias. Average annual precipitation in the trench is 46.9 cm (26.3 cm of rain and 205 cm of snow). This compares with measurements on the flanks of Mt Fidelity in the Selkirks of 218.1 cm of precipitation (48.9 cm of rain and 1692 of snow). Because of this dryness, the dominant trees are Douglas-fir and lodgepole pine as is normally found in the dry interior of B.C.

**Km 26.0:** The Columbia River below is about 200 km north of its source at Columbia Lake. The river at this point is flowing north in the trench and then loops around the Selkirk and Purcell Ranges to head south back down through Revelstoke and eventually into the USA.

**Km 47.4:** You are now cutting across the northern tip of the Purcell Ranges and into the Beaver Valley. The rock cut consists of slate, originally laid down as mud. Pressure and heat changed it to shale and finally to slate.

**Km 51.5:** McNaughton Lake, on the right in the distance, is a reservoir formed by the huge Mica dam near the apex of the "Big Bend" of the Columbia River.

**Km 53.1:** As you look up the Beaver Valley and to the west note the increase in clouds, greater diversity and density of the trees and underbrush, signs of a wetter climate.

**Km 54.8:** Entering Glacier National Park. Because of its ruggedness and inaccessibility, the park remains a wilderness and one of the last strongholds of the grizzly.

**Km 56.0:** The original Mountain Creek bridge across the valley was once the largest wooden structure on the CPR line. It was longer than the present steel bridge.

**Km 60.3:** Glacier Park gate.

**Km 61.0:** The Beaver Marshes and Beaver Valley are the best areas to observe wildlife in the park. Moose and beaver can occasionally be seen.

Beaver Valley divides the Selkirk Ranges from the Purcell Ranges. The Purcells are made up of less resistant rock than the Selkirks and hence are less rugged and lower than the Selkirks.

**Km 78.0:** Glacier Park Lodge. The Rogers Pass Centre building was patterned after the original railway snowsheds. The Centre has exhibits and displays about the human and natural history of the park. The Glacier Circle bookstore located in the Centre provides a large selection of books on the human and natural history of the area.

A 1.1 km leisurely walking trail runs from the Rogers Pass Centre to Rogers Pass Summit west of the hotel.

**Km 79.2**: Rogers Pass Summit Monument. The hiking trail from the Centre terminates here.

**Km 79.7:** There are more than 400 glaciers covering 155 sq. km of the park. Rogers Pass is ringed by alpine snowfields such as Asulkan Glacier ahead and slightly to the left. The largest, Illecillewaet Glacier, comes into view on the descent. This glacier used to extend down to the valley bottom, 1.6 km from the present highway. All glaciers in the park for which there are data are advancing. The Illicillewaet's net annual advance has averaged 10 metres since 1972.

**Km 82.3:** Illecillewaet Campground and Glacier House access road on the left on the outside of the 90° curve.

**Km 83.7:** Trail head for the Loop Brook trail. The hiking trail leads for 1.6 km past the huge stone piers built by the railroad over Loop Brook.

**Km 92.8:** An avalanche dam is on the left. Wet flowing avalanches often occur in spring and can reach 112 km/hr. Dry flowing avalanches that are more common in the winter can travel at 217 km/hr. Air borne powder avalanches can reach 257 km/hr. Impact pressures of 2,000 to 49,000 kg/ sq. metre have been recorded for flowing avalanches.

**Km 97.5:** Two avalanche gun positions on either side of the road are one of several such sites along the road. Avalanche guns are used in winter to knock out dangerous snow build ups above the road.

**Km 112.9:** The Albert Canyon Hot Springs are one of a number of natural hot springs found in the Columbian Mountains.

**Km 116.8:** Mount Revelstoke National Park.

**Km 117.8** Giant Cedars Picnic Site: This rest stop presents the rain forest of the Columbia Mountains. Western red-cedar, western hemlock, Douglas-fir and pine are found. Some of the cedars and hemlocks found at this stop and at the Skunk Cabbage site mentioned below, are 2.4 m across and 7.7 m around at waist height. Check the creek for a dipper. The trail that begins at the edge of the picnic site provides opportunities to spot several different bird species, particularly in June. In the odd marshy spot look for luxuriant skunk cabbage.

**Km 121.2:** Skunk Cabbage Site. The trail commences beyond the footbridge over the creek. The path loops through a variety of wetland habitats.

**Birds:** magnolia warbler, black-headed grosbeaks, rufous hummingbird, yellow warbler, Townsend's warbler and common yellowthroat. Several species of flycatchers call from the lowlands. Impressive skunk cabbages make the walk worthwhile.

**Km 145.4:** Mount Revelstoke Summit Rd Interchange. The only vehicle access into the central part of the park is a 26-km road to the summit of Mt. Revelstoke and the fantastic subalpine meadows found there. Late July though August is the best time. Meadows In The Sky Nature Trail is a pleasant one-half hour walk.

**Km 147.0:** The Revelstoke turn-off. The city of Revelstoke is also the administrative headquarters for both Glacier and Mount Revelstoke parks. Here the Columbia River has looped around the Selkirks and is heading south to the USA. The river separates the Selkirks from the Monashees. The wetland surrounding the Revelstoke Airport are the only extensive areas of marsh-like habitat left along the hundreds of kilometres of the now-dammed Columbia River. Numerous species of waterfowl use this as a staging area during migration (March-April, September-October). Rarities here include black-billed cuckoo, scissor-tailed flycatcher and LeConte's sparrow. The turnoff to Mica on Hwy 23 N offers some fabulous mountain scenery and a good place to see bears and caribou.

**Km 157.7:** Eagle pass summit cuts through the Monashee Range to the Columbia river valley. The pass also divides the Fraser and Columbia watersheds. The road now follows the Fraser water system to the coast.

**Km 166.3:** The rock cut shows a good example of the metamorphic rocks of the Monashees.

**Km 188.8:** Below here the CPR track was completed with the driving of the last spike on November 7, 1885. Here crews building from the west ran out of steel and waited for two months for the crews from the east to push through and link Canada from coast to coast.

**Km 199.5:** Yard Creek Provincial Park and picnic area.

**Km 218.0:** Sicamous and the junction with Hwy 97 that leads to the Okanagan Valley in the south.

**Further Information**

For further information, refer to the book *Glacier Country: A Guide to Mount Revelstoke and Glacier National Parks* by John G. Woods. • The field naturalists' group, Friends of Mt. Revelstoke and Glacier, is a useful contact group. • The Rogers Pass Centre, the major information centre in the region, is open year-round, and has plant and animal checklists and hikers' guides available.

*Keith Webb, and material from Canadian Parks Service*

Cedars will grow to immense size where there is a lot of moisture. Few of these big trees remain, as they are among the first to go when timber companies move in. ''Poster'' Big Cedar Tree in Mt. Revelstoke National Park.

Spider web made by *Aronews* sp..

**Page before:** Skunk cabbage *(Lysichiton americanum)* was used by native people to line food-steaming pits and wrap or cover food -- similar to the use of wax paper today.

Sitka deer fawn on Queen Charlotte Islands. This animal was introduced on to the Q.C.I. from Porcher and Pitt Islands.

Shaggy manes are found worldwide and grow in grass, lawns, farms, along roadsides and forests. Fruits in cool weather from August through October and later on the coast.

Northern alligator lizard *(Gerrhonotus coeruleus)* is widely distributed in southern B.C. from the East Kootenays to and on Vancouver Island. It is usually found in dry rocky areas where it can be seen basking in the warm sun or scuttling through dense cover. This lizard easily parts with its tail when frightened, but the operation is bloodless and a new tail soon grows from the old stump.

Glacier lily, *(Erythronium grandiflorum)*, blooms early in spring on subalpine meadows.

Pygmy owl feeding on starling near Revelstoke.

According to legend, the tridentate bracts of the Douglas-fir cone originated from little mice that scurried to hide in the cone when the Great Spirit was angry.

Sockeye salmon ascend most B.C. coastal streams to spawn in the fall. After hatching, the fry slowly move to the ocean where they mature, then return to spawn in the same stream in which they were born four or five years earlier.

Hoary marmot lives in the subalpine and is the largest member of the squirrel family, sometimes growing to twice the size of a woodchuck. In summer this marmot will put on a layer of fat accounting for 20% of its weight and then hibernate seven to eight months of the year. They live in loose colonies, taking advantage of neighbours' watchfulness for both bear species, wolves, and golden eagles.

**Facing page:** False solomon's seal, *(Smilacina racemosa)*, was classed with fairybells and twisted stalks as "grizzly-bear food" by the B.C. Interior native people who did not consider them edible. Other native people did eat them, but they are not too palatable.

Steller's jay is the provincial bird of B.C. It has similar habits and behaviour to its close relative the bluejay, but is less gregarious.

Tall bluebell, *(Mertensia paniculata)*, occurs in moist forests, woodlands, meadows and thickets into the supalpine. It is typically common in the north and in scattered places south to about 53 degrees.

**Next page:** Overlander Falls, Mount Robson Provincial Park.

Participants in outdoor bear safety course at grizzly bear den in the alpine near Whitewater Creek.

Rocky Mountain bighorn sheep have the largest horns of this group of mammals. Females of the species gather in bands of 15 or more for protection from predators, with the older females providing leadership. Rams older than age three, from spring until fall, band togrther in all-male groups of around a dozen which are led by older males. Golden eagles take many young lambs.

Highbush cranberry, *(Vibernum trilobum)*, provides brilliantly coloured leaves in fall. The red berries are fed on by birds in fall and early winter.

Gray jays are permament residents of coniferous and mixed wood forests. They nest in mid winter by building a very thickly lined nest using hair, fine grass, moss and sometimes coniferous needles, and incubate the eggs and nestling at all times.

**Next page:** Fraser Lake at the end of a good day, west of Vanderhoof. **inset:** Northwestern toads, *(Bufo boreas)*, found throughout the province, congregate in early spring in swamps and along margins of ponds and lakes to spawn. At that time males call with a high-pitched tremulous note, amplified by the vocal sacs distended beneath the skin.

Rufous hummingbird males have a distinctive U-shaped courtship dive performed each spring. Once courtship is over, males depart leaving females to raise the young. These hummingbirds are attracted to brightly coloured flowers and often come to feeders set up in back yards from early spring to late summer.

Pacific dogwood, *(Cornus nutralli)*, is the floral emblem of B.C.

Fall colours, poplar plus others.

Northern saw-whet owls nest in abandoned woodpecker holes, with males each spring for a few weeks giving a whistled song of regular toots which some people say sounds like a saw being sharpened.

**Next page:** The seed heads of western anemone, *(Anemone occidentalis)*, are sometimes called towhead babies.

# CENTRAL BRITISH COLUMBIA

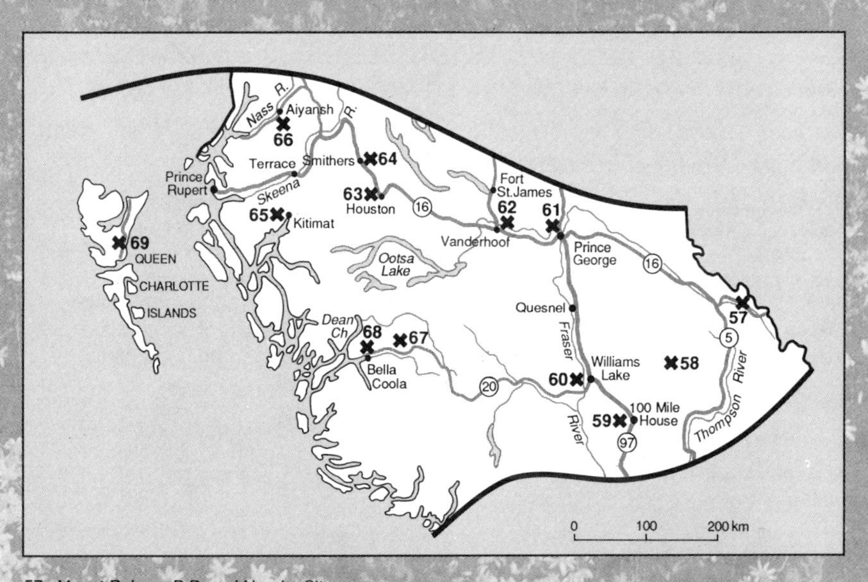


57. Mount Robson P.P. and Nearby Sites
58. Wells Gray P.P. and South Areas
59. 100 Mile House and Area
60. Williams Lake
61. Prince George
62. Vanderhoof
63. Houston - Topley Area
64. Smithers Region
65. Kitimat
66. Aiyansh (Nass) Volcanics
67. Tweedsmuir P.P.
68. Bella Coola
69. Queen Charlotte Islands

# MOUNT ROBSON PROVINCIAL PARK AND NEARBY SITES

Lying against Jasper National Park, Mount Robson contains a wide variety of birds, mammals and flowering plants similar to, but different from, Jasper.

The park contains the headwaters of one of B.C.'s longest and most important rivers, the Fraser, and the highest Canadian Rockies peak, Mt Robson at 3,954 m. Berg Glacier, in the northwest section of the park, is one of the few advancing glaciers in the Canadian Rockies. Huge chunks of ice fall into Berg Lake each year. A park trail to this lake begins from a parking lot beside Robson River two km from Hwy 16. It is about a two-day hike in and out.

Between Moose and Yellowhead Lakes there are several marshy areas, ponds and muskegs where birds are abundant in numbers and species.

**Birds**: Thrushes, flycatchers, vireos, warblers, ducks, geese, Clark's nutcrackers, black and Vaux swifts, robins, cedar waxwings, pine siskins, kestrels and sharp-shinned hawks.
**Mammals**: Moose, mountain goat, grizzly bears (north of Yellowhead Lake), marmot, pika, beaver, muskrat, mule deer, black bear, red squirrel, elk, caribou and lynx.
**Fish**: Dolly Varden, lake char, kokanee and rainbow trout.
**Plants**: Lodgepole pine, birch, spruce, Western red-cedar, subalpine fir, alder, white bark pine, black spruce, red-osier dogwood, thimble berry, soopolallie, and huckleberry.

### Overlander Falls

The trail down to the falls passes through a good mix of vegetation. The sign and trail lie along Hwy 16, 1.5 km east of the main Robson parking lot and viewpoint.

### Rearguard Falls

This spot on the Fraser River is the final barrier to chinook salmon migrating up-river from the Pacific ocean about 1,200 km away. In August they can be seen attempting to jump the 10 m stepped falls. The site, along Hwy 16, sits 12.5 km west of the main parking lot and interpretive centre, and about 4.5 km east of the Tete Jaune Junction.

### Teare (McBride) Mountain

Alpine flowers are lush in the alpine meadows above the forestry lookout station that sits on the top of this mountain. The difficult dirt road (mud after rain) is only accessible by four wheel drive in summer. To reach the site drive to the bridge over the Fraser River at McBride, west of the park on Hwy 16 and 63 km west of the Tete Jaune junction. Return east from this bridge on 16 for about a km and take the first left road, Mountainview, and follow it around for about 0.2 km. This road is very near a house and looks like a driveway. Watch for a road to the north (right) turn off, which is the road up the mountain. After about another 0.2 km take the left fork.

*David Stirling, and* The B.C. Parks Explorer *by Maggie Paquet.*

# *WELLS GRAY PROVINCIAL PARK AND SOUTH AREAS*

Wells Gray is the third largest park in B.C. The vulcanism of southern Wells Gray is the most fascinating single feature to explore. Volcanoes here have been emptying off and on over the last half million years, often at the same time as glaciers covered the land.

The northern two-thirds of the park has no volcanic rocks; instead this region is made up of metamorphic rocks including gneiss and schist.

Most of the park lies within the interior wet belt. Summers are warm with moderate rainfall and winters are cool, often with heavy snowfall. The three biogeoclimatic zones found in the park include Interior Cedar/Hemlock, Englemann Spruce/Subalpine Fir, and Alpine Tundra.

The park is also good for birding with 218 species recorded, at least 82 of which breed locally. Trophy Meadows with two waves of flowers from early July 1 through early to mid-August, provide one of the best floral displays in B.C. Visitors should begin a visit at the Green Mountain Viewing Tower, accessible by vehicle from 0.4 km inside the park gate.

## North Thompson Provincial Park

This was a home of the Nsimpxemux people and one of the major Shuswap Indian settlements along the Thompson River.

Bird watching is good at the lookout at the top of the river trail. A great blue heron rookery lies nearby.

The park lies 5 km south of the community of Clearwater, on the east side of Hwy 5.

**Birds**: herons, ducks, geese and bald eagles.
**Mammals**: deer and many small mammals.

## Yellowhead Museum

The museum is open to the public, at no charge, from June 15 to September 15, and by appointment from May 1 until Oct 31. It lies near the Wells Gray Rd, 7.2 km north of the Yellowhead Hwy 5 turnoff north towards Wells Gray park. About 6 km up this road a sign on the right (east) directs you up the 1.2 km gravel road to the site.

## Spahats Creek Provincial Park

Spahats (Indian for bear) Creek has carved a 122-metre deep gorge through a series of lava flows. A viewpoint only a few metres from the parking lot provides a vista of the many layers of these flows. Here you also get a look at the 61 m drop of Spahats Creek tumbling over the volcanic precipice into the Clearwater River far below. At this site there are a series of geological displays illustrating the volcanic history of Well Gray Park.

The park lies just off the Wells Gray Rd, about 15 km north of the Clearwater turnoff from Hwy 5. From the campground turnoff, continue about a km north along a gravel road to an excellent viewpoint for the vista up the Clearwater valley.

## Trophy Meadows

Northern and mountain birds are abundant here, but the main attraction of this area is the flower meadows that stretch across the Trophy Mountains, blooming from July 1 into early or mid-August.

To reach the Trophies, follow the Upper Clearwater Valley Rd for 11.5 km north of Clearwater to a point just north of Spahats Park, turn right onto a gravel road signposted "Trophy Mountain Recreation Reserve." Follow the signs over the next 15 km, as you climb to about the 1750 metre level. From here the Trophy meadows are a leisurely hour away. Hiking distance from the parking area is about 4 km and elevation gain is 250 metres.

**Birds**: Boreal chickadees, blue grouse, rosy finches, golden eagle, spruce grouse, white-tailed ptarmigan, hermit thrush and other woodland species.

## Helmcken Falls and Dawson Falls

Dawson Falls cascades down 18 metres. A short trail leads to several vantage points overlooking these falls. The trail begins 4 km north of the Hemp Creek entrance.

The Mush Bowl or Devil's Punch Bowl lies a little further along the main road where it crosses the Murtle River. Here the river has poured through the narrow gorge for ages, using boulders to grind huge holes in the rock.

Nearly three times the height of Niagara Falls, Helmcken Falls at 140 metres overwhelms a visitor arriving at the viewpoint. It was apparently created when the Murtle River, having carved a canyon deep into Murtle Plateau, suddenly changed its course to enter the canyon from a new direction, and hence the falls. Each winter an ice cone of from 40 to 60 metres high forms around the base of the falls. Over time such cones have excavated an immense cavern into the canyon walls. To reach these falls take the road to the left (west) just north of the Dawson Falls Campground and Falls, just over 4 km north of the Hemp Creek entrance on the road north from Clearwater.

## Ray Farm

This is the region's hot spot for bird and mammal watching. In June or July more than 50 species of birds can be seen.

This site is one of three locations in B.C. to find adder's tongue fern, and white rein orchids grow in mass profusion. Another feature is the mineral springs. They provide a rich source of iron, sodium, calcium, and magnesium. Alice Lake is known for its water lily display in July.

The walk from the park road into the homestead site takes about 10 minutes.

**Birds**: Common snipe, veery, white-winged crossbills, red-tailed hawks and northern waterthrush.
**Mammals**: Mule deer, moose, black bear, Columbian ground squirrels and coyotes.

## Clearwater Lake Campsite and the Dragon's Tongue

The campsite sits on a distal end of a 16-km-long lava flow, the Dragon's tongue. This flow crossed the Clearwater Valley damming the lake and raising the water three metres above its original level. The river has since cut into the lava, creating Osprey Falls near the campsite. In some areas, for example 2 km up the Kostal Lake trail, the surface of the Dragon's Tongue consists of broken blocky rock, massed into ridges and troughs two to five metres deep. Tree frogs live in this inhospitable environment, and apparently are able to satisfy their moisture requirements in the lava crevices.

**Birds**: Vireos, warblers including MacGillivray's, osprey, and dipper.

From the campground, to reach the main lava body, a marked trail leads 2 km to more exposed portions of the tongue. The trail passes the Dragon's Den, a grotto worn into the side of the flow. Later can be seen the Dragon's Teeth, vertical and diagonal wells in the lava where tree trunks once stood as part of a forest that was inundated by lava. Check the inside of these wells to find bark impressions. The campground lies about 72 km north of Clearwater.

## Murtle Lake

Lying on the east side of Wells Gray, in the Nature Conservancy Area of the park, this 40-km-long lake is open only to boats with paddles — no motors allowed.

Near the top of Central Mountain — a seven-hour round-trip hike — mountain caribou are frequently seen lounging on late summer snow patches.

**Birds**: Osprey, eagles (nesting), warblers including blackpolls, which is rare in the region, three-toed woodpeckers, spruce grouse, chestnut-backed chickadees, loons and Clark's nutcrackers.
**Plants**: whitebark pine,

The lake is reached from Hwy 16 at the Blue River turn where the sign points to Wells Gray Park. Take the 25 km gravel mountain road which is narrow in places. It terminates at the park boundary where a wide level trail provides an easy portage of 2.5 km to the lake launch site. About a dozen campsites are available for canoeists along the shores of Murtle Lake.

### Further Information

There are no services within the park, so gas up at Clearwater. • The Wells Gray Visitor Centre, located at the junction of the Yellowhead Hwy. and the Wells Gray Provincial Park Rd, offers a full compliment of up-to-date information on Wells

Gray. Be sure to pick up a copy of the Wells Gray bird checklist and "Nature Wells Gray," both written by Trevor Goward and published by the Friends of Wells Gray, Box 1386, Kamloops, B.C. V2C 6L7. For detailed information contact the Zone Supervisor at Park Headquarters, Box 70, Clearwater, B.C. V0E 1N0. The Park has several handouts, including maps for canoeing and cross-country skiing. Write for them. The topographic map is sheet NTS Map #82M/13. • Exploring Wells Gray Park by Roland. Neave, 1988, published by The Friends of Wells Gray Park, is a good reference and trail guide.

*Trevor Goward, Ida DeKelver and Barbara McCuish*

# 100 MILE HOUSE AND AREA

The area offers a nesting colony of herring gulls (the only one known for the region) and black swifts at Bridge Lake, and a marsh in downtown 100 Mile House that is being upgraded by Ducks Unlimited. At Canim Lake look for a variety of nesting birds.

The 108 Ranch and the Sucker Lake Rd are excellent for wildflowers. Nearly 120 species have been noted in the area including shooting star and shrubby beard's tongue.

**Birds**: four grebe species, American bittern, teal, mallard, ruddy duck, sora, coot, snipe, great grey owl, red-naped sapsucker, willow flycatcher, swallow (tree, violet-green, rough-winged, cliff and barn), warblers (Nashville, yellow, Townsend's, northern yellow throat, Macgillivray's and Wilson's), Savannah and Lincoln sparrows, red-winged and yellow-winged blackbirds are regular breeders.
**Mammals**: beaver and muskrats.
**Plants**: lambs quarters, foxtail barley, some giant rye grass, and clover, willows, spruce, aspen, red-osier dogwood, rose, black twinberry honeysuckle, soopalallie (buffalo berry), waxberry (snowberry), bullrushes and juncus.

## 100 Mile House Marsh (aka Sanctuary, Refuge)

This 32.5 acre wetland is fed by underground springs.

Mid April to the end of May and September through October are the best times to see migrants including tundra swan, both yellowlegs, Wilson's phalarope, herring, ringed-billed and mew gulls (the herring gulls come from the breeding colony on Gull Island on nearby Bridge Lake). Reclamation of the marsh will hopefully reestablish the black tern as a regular breeder.

Current access to the trail is from the airport on the north. Future access will be from the south. Benches have been placed along the route. Three-quarters of the marsh is readily visible from most of the accessible area. The walking trail lies on the south side.

Look for the marsh on west side of Hwy 97 in the middle of the community, at the Tourist booth.

## Bridge Lake

This large picturesque lake has several islands. One, Gull Island, is a small bare rock outcrop and contains a herring gull colony with 20 to 30 nesting pairs of birds. It lies no more than 500 m from the southeast shore.

Access to Gull Island is by boat, but it should be avoided during the breeding season as the birds will eat each other's chicks if disturbed. It is best to stay at least 250 m away from the island and watch the gulls through binoculars or spotting scope.

The lake lies along Hwy 24, equidistant (48 km) from Hwy 97 at the turnoff at 94 mile House, 9.6 km south of 100 mile House, and Little Fort which is 96 km north of Kamloops on Hwy 5 (Yellowhead). Hwy 24 is a fully paved road connecting Hwy 97 North at 94 Mile House in the west and Hwy 5 at Little Fort in the east.

## Canim Lake

The area around the lake is about 90m lower in elevation than at 100 Mile House and thus is slightly warmer and drier. This accounts for the small colony of bobolinks that are found adjacent to the Canim Lake Resort at the south end of the lake. Land around the resort has aspen, willows and cattails and is the best place to birdwatch. American bittern nest in the cattails near the resort and a pair of bald eagles occur nearby with a very visible nest.

This large lake lies about 35 to 40 km north east of 100 Mile House, on a good paved road on the way to the south east corner of Wells Gray Provincial Park. To reach the lake, leave Hwy 97 about one km north of 100 Mile House, and follow the road north east through Buffalo Creek to Forest Grove and then east through the Canim Lake Indian Reservation. No provincial campsites are found along the road, but there are numerous private locations.

A visit to Mahood Falls near Canim Lake is worth the trip.

## 108 Mile Ranch

This is the best spot for wildflowers in the region. They begin blooming in April with buttercups coming up between melting chunks of snow. Various species bloom until fall freeze up in September. The best months to look for flowers are June and July. Take Sucker Laker Rd opposite the Heritage building site to see the flowers. The "look-out" which lies on the road to the right off Sucker Lake Rd has a good variety of flowers such as shooting stars and shrubby beard's tongue that are found nowhere else in this area. This "look-out" lies on private property, so be respectful. The owners ask that people leave the site as they find it, so take out all garbage, and don't drive off the road.

**Further Information**

100 Mile House lies 115 km or 1.5 hours north of Cache Creek on Hwy 97. Provincial park camping is available at Loon Lake, Big Bar Lake, Bridge Lake and Lac La Hache 25 km north of 100 Mile House. A Tourist Bureau, open from May 1 to the September long weekend, lies in 100 Mile house on Hwy 97 at the corner of 4th St near the Bird Sanctuary.

*Laurie Rockwell and Angela Skeene*

# WILLIAMS LAKE

Lying within the Cariboo ranching and forestry country, the region has a wide selection of natural history sites to explore. Over 250 species of birds have been recorded for the area, as it lies on the Pacific flyway and contains upland grasslands and forests with numerous lakes.

## Scout Island and Nature Centre

The site, within the city limits and at the western end of Williams Lake, offers opportunities to watch wildlife on the adjacent lake and marsh. A nature house, open May-September, provides natural history displays and information for visitors. Walking trails provide opportunities to explore the area. A bird check list can be picked up at the Nature House. In summer, staff provide nature programs for children.

The Nature Centre is only a km off Hwy 97.

## Alkali Lake

Found straight south of Williams Lake and Spring House, Alkali Lake was set aside by the B.C. Ministry of the Environment as a waterfowl sanctuary. This is an excellent bird watching area, particularly in spring.

## Farwell Canyon

Magnificent badland hoodoo formations and sand dunes are found here. California bighorn sheep sometimes may be seen.

To locate the canyon drive about three-quarters of an hour west of Williams Lake on Chilcotin Hwy 20. A sign at Riske Creek points south through rangeland. Follow this road until it drops into the Chilcotin valley through a series of switchbacks to cross the Chilcotin River at Farwell Canyon.

## California Bighorn Junction Sheep Reserve

The reserve holds the northernmost band of California bighorn sheep in the world. Transplants are made from this healthy herd to repopulate other areas in B.C. and the northern U.S.A. In July and August sheep are often visible from the road. During the breeding season in late September through October, large numbers gather on flat plateaus along the Chilcotin River.

The Reserve is located on a triangle of land formed by the junction of the Chilcotin and Fraser Rivers, south of Riske Creek. To find the site, drive a short distance north of Farwell Canyon (see above), and take the back country trail to the east.

### Chilanko Marsh

Excellent bird watching for waterfowl is found here. Shorebirds and raptors come in fall and spring. The marsh lies 177 km west from Williams Lake on Hwy 20, close to Chilanko Forks.

### Horsefly - Quesnel Lakes

Horsefly Lake, to the east of Williams Lake, holds large lake and rainbow trout. Beyond this lake, 36 km of scenic road leads through a wilderness of small secluded lakes. Quesnel Lake, at the end of the road, contains trophy Rainbow and lake trout, Dolly varden and kokanee salmon.

Near Quesnel Lake huge stands of old-growth cedar and hemlock are found. Hiking trails lead to alpine meadows full of flowers in July.

Another 12 km beyond Horsefly Lake, is the trailhead for Viewland Mountain Hiking Trail. This well marked 3 km long trail winds casually up an altitude gain of 400 m. The path is enclosed by huge red-cedars almost to the top where it enters alpine vegetation.

**Birds**: mountain bluebirds, western meadowlarks, and black-capped chickadees, ruby-crowned kinglets, Steller's jays, and woodpeckers.
**Mammals**: mule deer, coyote, fox, black bear and moose.
**Plants:** lodge pole pine, white/Englemann hybrid spruce, saskatoon, blueberries, bunchberry, twinflowers and thimbleberry.

**Further Information**
Williams Lake is on Hwy 97 north of Cache Creek and south of Prince George.

*Frances Vyse, Anna Roberts and Fred McMechan*

# PRINCE GEORGE

The area abounds in birds, particularly in May and June. Trumpeter swans winter on the Crooked River north of the city. Moose and Mule deer are abundant.

Prior to the last ice age, rivers flowed north from near Williams Lake and northeast into the Peace River to the Arctic. After the glaciers melted away, about 10,000 years ago, a huge shallow lake filled the Prince George basin with its melt waters. Runoffs from the retreating ice and from surrounding high country poured into the lake for hundreds of years bringing in gravel, sand and clay to be deposited in layers that now total over 100 metres. Then as the land rose with the further retreat of the glaciers, the two rivers, the Fraser and

Nechako, cut through these layers leaving the cutbanks and Connaught Hill as reminders of these geologic processes.

### Cottonwood Island Nature Park

No longer an island, this Nature Park has 3.5 km of trails under a stand of cottonwood trees; some were present when Alexander Mackenzie went through nearly 200 years ago. Understory consists of a variety of Central Interior shrubs including hawthorn and high-bush cranberry. The varied vegetation provides good birding all year with exceptional opportunities during migration. Interpretive signs and trail leaflets provide information about the natural history.

A new Heritage Trail System extends from Cottonwood Island Park westward up the Nechako to the old wooden bridge, and goes southward from Cottonwood Island along the Fraser to South Fort George Park and Museum, and then along Hudson's Bay Slough.

The Park is reached by taking First Ave towards the Fraser River. Then, using the interchange, drive over the railway and turn right at the first road. Drive to the parking lot or continue down River Rd to a second parking lot at the west end of the park.

### Moore's Meadow Park

Wooded slopes enclose a depression about 1.5 km long with trails running throughout. Over 100 native flowers have been found blooming here. Old trails provide easy walking in summer and good cross-country skiing in winter.

### Wilkins Regional Park

The park contains a variety of habitat along the Nechako River. Bird life is abundant from early spring to fall. A nature trail, for which there is a brochure, passes through wet areas of willow, alder and birch and higher lands with lodgepole pine and spruce. To reach the park take Otway Rd for 14 km to Miworth Rd. where you turn south for a very short distance and then west on Wilkins Rd into the park.

### Moose Viewing Station

There is a nature trail loop at this site which lies along Hwy 16 east, about 1.5 km east of the Tabor Mountain Ski Lodge turnoff, which in turn is about a 20 to 30 minute drive east from Prince George.

### Fort George Canyon Trail

This 4.5 km (one-way) trail passes through groves of Douglas-fir, lodgepole pine, aspen, spruce and birch. Look for an amazing diversity and quantity of flowers, shrubs, grasses, ferns, mosses, and lichens. Over 50 species of flowers have been seen, together with 24 shrub species and 13 species of trees.

**Birds**: varied thrushes, goshawks, red-tailed hawks, pileated woodpeckers, bald eagles, pine siskins, redpolls, pine grosbeaks and crossbills (winter).

**Mammals**: mule deer, black bear, moose, coyote, fox, mice, red squirrels.
**Plants**: pine drops, Indian pipe, green pyrola, one-sided pyrola, large round-leaved orchid, water avens.

There is an excellent pamphlet describing the vegetation at self guiding posts along this trail.

The canyon is a formidable sight. To reach it, go west from the Tourist Bureau at the junction of Hwys 16 and 97, and take Hwy 16 for 11 km watching for the Blackwater Rd turnoff. Go left (south) on this road for

11.3 km to the West Lake Rd. Follow West Lake Rd for 9.5 km past several side roads to a major fork. Take the left fork for 1.8 km before turning into a narrow old track. Drive this track for 1.8 km to a small parking lot. The trail is level to the lookout followed by several descents which lead down 180 metres.

## Berman Lake Regional Park

Lying beside Berman Lake, the park offers good bird watching opportunities. It is just off Hwy 16, 45 km west of Prince George. Watch for a sign on the south side of the highway.

## Tamarack Lake

The small stand of tamarack at this lake is near the southern limit for this species. The grove of trees is visible from Hwy 16, about 48 km east of Vanderhoof, just west of the lake on the north side of the highway.

## Cinema Bog Ecological Reserve

The reserve preserves an outstanding example of a northern lowland sphagnum bog near the southern limits of this vegetation type. There is an exceptional diversity of sedges (11 species) in this one small area. A typical northern interior spruce forest margin must be passed through to gain access to the bog proper.

**Plants**: Labrador tea, bog rosemary, sphagnum moss, wetter areas scrub birch, mosses, sedges, cloudberry, bog cranberry, moorwort, cotton grass, and long- and round-leaved sundews.

Access is about 88 km south of the new Fraser River Bridge on Hwy 97 at Prince George. Look for the low trough-like area on your left a few km south of Dunkley's sawmill. Due to the sensitive nature of the site, it is requested that visitors contact the Prince George Naturalist club and the volunteer wardens who reside in the area for access details. They have other Ecological Reserves near Prince George that also could be visited. The local Chamber of Commerce has the address and phone numbers.

## Crooked River

This relatively slow moving river runs parallel to Hwy 97, north of Prince George. The valley lies on the north-south flyway over a low part of the Arctic Divide and thus the more than 105 species of birds that have been recorded in Crooked River Park are only part of the bird life to be seen along this waterway. Springs along the river help to keep some of it open all year. Trumpeter swans and other waterfowl use the river in the winter together with bald eagles. Moose winter along the river with one January survey recording 18 bulls, 32 cows and 14 calves.

**Mammals**: moose, a variety of mice, black and grizzly bear, wolf and red fox.
**Fish**: three species of sucker, squawfish, chub, Rocky Mountain whitefish, rainbow trout ,Dolly Varden, burbot, lake char and lake whitefish.
**Reptiles and amphibians**: common garter snake, northwestern toad, wood frog and spotted frog.
**Plants**: willow, alder, lodgepole pine, white spruce and balsam fir.

The Teapot, a bump rising about 245 m above its surroundings, has a steep trail to the top with a circular loop around the summit. To reach this spot drive north on Hwy 97 beyond the Summit Lake area and turn off on Tallus Rd. Turn right on the road to the Forest Recreation Site and drive on to the bridge over the Crooked River. Just across the river look for the parking area and signs pointing to the start of the Trail.

**Further Information**
The Fraser-Fort George Museum lies on 20th St off Queensway; phone 562-1612. The Recreation Officer for the Prince George Forest Region has information on most of these sites, at 1011 - 4th Ave Prince George, phone 565-6100. Brochures with maps of most of the parks including the nature trail brochures are available from the tourist bureau at the corner of Patricia Blvd. and Victoria St. The B.C. Forest Service has a good collection of Recreational Site maps that cover most of the area. These maps may be obtained free from the 5th floor, Government Building, the same address as for the Recreation Office. A Prince George and District Trail Guide is published by the Caledonia Ramblers, a local hiking group. The booklet outlines about 25 hikes in the area including several right around the city plus some as far as Barkerville in the south, Valemount and McBride to the east, Fort St James to the north west, and Mackenzie and Pine Pass to the north.

*Mary Fallis, Marilyn Rack and Don Wilson*

# VANDERHOOF

Canada geese gather in the thousands within the community of Vanderhoof as the birds move north in April and south in autumn. Most of the other good spots lie west and south of this community.

## Federal Migratory Bird Sanctuary

This natural widening of the Nechako River with its maze of small islands and gravel bars within the town of Vanderhoof, has been the stop-over resting and feeding place for huge flocks of Canada geese since time immemorial. The site, the only sanctuary of its kind in north Central B.C., lies on the Pacific Flyway migration route. These river flats in April/early May, together with fields adjacent to nearby roads and the highway, are covered with geese. In fall, smaller but still significant numbers of birds are present. Migratory hawks and owls are also common as they move through along the flyway.

The spot is accessed from the north end of Lampitt Ave, but the bulk of the sanctuary lies upstream from the Nechako Bridge.

## Beaumont Provincial Park

Located at the eastern end of 19-km-long Fraser Lake. The park lies on the upland Nechako Plateau.

**Birds**: common loon, ruffed and spruce grouse, mallard, northern pintail, rufous hummingbird, several warblers including orange-crowned, Wilson's and yellow-rumped, and savannah sparrow. Mammals: moose, mule deer, and black bear.
**Fish**: rainbow trout, burbot (ling cod), sturgeon, kokanee salmon, lake trout, and char.
**Plants**: birch, poplar, spruce, willow, aspen, saskatoon, raspberry, black huckleberry, high-bush cranberry and blueberry.

The park entrance lies on Hwy 16, 4 km west of the community of Fort Fraser and 44 km west of Vanderhoof. There are 41 campsites here.

## Nautley River

The Nautley, at one km, is one of the shortest rivers in the province. Trumpeter swans over winter here. Sockeye salmon are seen as they move enroute to spawn at Francois Lake/Stellako River (see below). Salmon are best watched at the Nautley bridge from July through October.

To reach the site drive north from Hwy 16 at Beaumont Park, along the Nautley Rd to the bridge across the Nautley River.

## Drywilliam Lake and nearby Ecological Reserve No 60

Good examples of interior Douglas-fir are visible on rocky outcrops towards the west end of Drywilliam Lake. The Ecological Reserve lies between Fraser Lake and Hwy 16, just past the west end of Drywilliam which is on the south side of the highway. Access to the reserve is poor but part of the fir stand is visible on the north side of Hwy 16 opposite the Orange Valley Motel.

## Ellis Island

This low rocky island in east central Fraser Lake, just west of Drywilliam Lake, is a ringed-bill gull rookery, only one of two such rookeries in B.C. The other nesting site lies on Grant Island in Okanagan Lake. Loons and ducks nest on the island too.

**Plants:** trembling aspen, black cottonwood, alder, willow, gooseberry, rose, raspberry, red osier dogwood and stonecrop.

Access is by boat from Beaumont Provincial Park on the east side of the lake, or at the town of Fraser Lake at the west end of the lake of the same name.

## Stellako River/Francois Lake

This is a staging area for trumpeter swans. The site occurs adjacent to where Hwy 16 crosses over the Stellako River. As soon as open water appears in the spring the birds arrive. The seven-mile Stellako River is a spawning site for sockeye salmon starting about the second week in September. A spacious viewing area lies about a half mile from the highway. Spawning sockeye salmon are visible from the Glenannan Bridge in the fall. The shores of Francois Lake are popular with rockhounds who find a type of banded agate with reddish to pinkish hues occurring in nodules (thunder eggs) along the beach and in rock cuts.

The drive on the gravel road that runs along the north shore of Francois Lake brings you through Douglas-fir stands and park-like grasslands on the south-facing slopes. The rockier outcrops support some Rocky Mountain juniper shrubs near their northern limit. Deer are often spotted on this drive.

The site lies on Hwy 16 at the west end of Fraser Lake, where the road crosses the bridge. The road to the north shore of Francois Lake takes off from the highway just east of the bridge. Another road, also to the north shore of Francois Lake, heads south just east of the community of Endako.

## Cheslatta Falls/Nechako River

The upper Nechako River is relatively isolated and thus wildlife is often present along the banks as you canoe down the river.

**Birds**: bald eagles, osprey, kingfisher, harlequin duck, mergansers and Canada geese.
**Mammals**: moose, deer, black bear, grizzly (occasional), beaver, otter, coyote and wolf.
**Trees**: white spruce, lodgepole pine, trembling aspen, black cottonwood, and birch.

Canoeing all, or part, of the Nechako can be a pleasurable experience. The total length of the river is approximately 247 km. Cheslatta Falls to Fort Fraser is 77 km; Fort Fraser to Vanderhoof is another 48 km; and it is 122 km from Vanderhoof to Prince George. The first 15 km from Cheslatta Falls downstream are Grades 1 to 2 with three moderate rapids. Then there are small canyons with rapids at 42 km and 28 km from Cheslatta Falls. These two sets can be quite severe at low water. Three more rapids are found between Fort Fraser and Vanderhoof, and three others between Vanderhoof and Prince George.

Canoes may be launched at Cheslatta Falls at the east end of the lake of the same name. This site is reached by driving south off Hwy 16, 8 km west of Beaumont Provincial Park, on the Holy Cross Forest Rd.

Proceed south for about 15 km and then turn right (southwest) and follow this road for about another 35 km to the falls and river. Canoes may also be launched at Fort Fraser and Vanderhoof plus other spots along this river. There is a forest service campsite at Cheslatta Falls.

### Grand Canyon of the Nechako

This 8-km, dried-up canyon provides views of water-carved rocks, pinnacles and caves, sinkholes and water-filled potholes with some of the grinding boulders still present. Towering overhanging cliffs are found in some sections and others have huge boulders.

The best access to the canyon is from Cheslatta Falls where you can walk up the dry river bed. Access into the canyon itself is fairly difficult and dangerous with sheer rock walls and large pools of standing water.

**Further Information**
Vanderhoof lies west of Prince George along the Yellowhead Hwy 16. Additional help is available at the Vanderhoof Museum, and Village Municipal Office.

*Nechako Neyenkut Society*

## HOUSTON - TOPLEY AREA

The region abounds with wildlife, including moose. Several creeks have annual spawning runs of salmon and spawning lamprey may be watched at Lamprey Creek. Alpine flowers grace the mountain tops. The area is riddled with forestry and mining roads, many of which may be used as good access routes into the wildlands.

The Morice River valley, south of Houston, is a major migratory corridor for birds in spring and fall. The valley also is used as a corridor for moose as they move from the uplands to winter range in the lowlands. Mountain goats are often seen on slopes of nearby alpine ridges.

**Birds**: bald eagles (Francois Lake), swans, golden eagles (Morice Lake) and waterfowl.
**Fish**: steelhead trout, spring, coho and pink salmon, rainbow, dolly varden and cutthroat trout.

### Babine Lake Area

Salmon have a major spawning site on the Babine River near where the Forestry Rd crosses the river just north of Nilkitkwa Lake, which in turn is at the north end of Babine Lake.

The lower end of Fulton River and the northwest end of Fulton Lake support a variety of waterfowl in September-October. Bear, moose and deer also may be seen here. The east end of Fulton River has spawning enhancement channels where the fish come each fall. To reach Fulton Lake, take the paved road north from Topley and turn west at Km 43 on to the forestry road, and drive 8 to 10 km southwest to the lake. The spawning channels on Fulton River are reached at Km 39 on the same highway.

Nearby Red Bluff Provincial Park has iron-stained cliffs that plunge dramatically into Babine Lake. The park has campsites, a beach and a boat launching site.

Pinkut Creek, north of Burns Lake, and just south of the major bend in Babine Lake, also has salmon spawning channels as part of the enhancement program.

### Francois Lake

The free ferry across Francois Lake, straight south of the community of Burns Lake along a paved road, offers chances to see swans that winter or rest on their way through in the spring, together with other waterfowl.

The west end of Francois Lake, south from Houston using the Morice River Rd, hosts numerous waterfowl of several species in September and October. Bald eagles gather here in May and June. Close to 50 eagles have been seen on the bluffs at the west end of the lake, west of Noralee West Recreation Site. River otter are spotted regularly by boaters near the shores of Francois Lake.

### South of Houston

Heading south from Houston on the Morice River Rd, the Swiss Fire area of 18,000 hectares found near the junction of the Morice and Owen Rivers, supports numerous woodpeckers and other birds all summer. Moose, deer and black bear also may be observed along the Morice River Rd between Km 18 and 28.

East of Morice River and south from Houston along the Buck Creek Rd between Km 18 and 28, on Buck Flats, is another spot to watch for the large mammals.

Back on the Morice River Rd, Lamprey Creek, located at Km 44, is a good spot to watch lamprey spawn between April and May.

Spawning lamprey also may be seen up Owen Creek at the west end of the Owen Lake Recreation site. This spot lies a few km south of the Swiss Fire area mentioned above.

The outlet of Morice Lake into Morice River is a good spot to seen golden eagles. Most of the Morice chinook salmon run in the upper reaches of Morice River immediately below the outlet of Morice Lake. This river also is recognized for its world-renowned steelhead run.

Nearby McBride Lake to the east is a good spot to see bear on the slopes to the north of the lake.

### Sweeney Mountain

The alpine areas on this mountain support an abundance of flowers including forget-me-not and lupin in late July and August. In the same area hoary marmots in a relatively large colony (12-15) gather vegetation for their nests in September. *Note*: Access to the mountain is by 4-wheel drive. Head south of Houston on the Morice River Rd, but continue south when this road swings west and continue on beyond Owen Lake, mentioned above, along the Nadina River to the Nadina Spawning Channel area.

Continue south past Twinkle Lake; from this lake you need a 4-wheel drive. You will pass through a mass of huckleberries south of Sibola Creek; watch for bears. The road turns west and passes through an old gravel pit. Just west of Sweeney Lake, turn north on an old mining road up Sweeney Mountain. The road is quite steep, with four or five switch backs, all the way to the alpine site. The drive up is worth the view of masses of blooms in late summer.

**Further Information**

Houston and area lie along Hwy 16 west of Vanderhoof. Accommodation and other services are available in Burns Lake and Houston. The area is riddled with forestry and old mining roads. They are often in relatively good shape, but check with the forest service office in Houston before leaving the highway. Good maps showing the roads, lakes, creeks, mountain peaks and Forest Service Recreation Sites are available from the headquarters of the Morice Forest District. Before heading into the back country, it is wise to at least check in with them. Phone at 845-6200, or write P.O. Bag 2000, Houston, B.C. V0J 1Z0

# SMITHERS REGION

## Bulkley Valley

This area, different from most other B.C. natural history sites, consists of farmland surrounded on three sides by spectacular mountains.

The climate in the valley is basically continental. Moisture laden winds from the coast usually drop their precipitation over the coastal mountains before reaching the valley, causing relatively dry weather. However, not all the moisture is lost over the coastal ranges since the westerly side valleys and the lower reaches of the main valley north of Moricetown have more coast-like vegetation.

A wide variety of habitats and many small lakes mean that bird watching is good in spring and summer. Many small landbirds, shorebirds, waterfowl, owls and hawks nest in the valley.

Valley vegetation is a complex mix of mature coniferous forest dominated by hybrid spruce (*Picea glauca* X *Engelmanni*) and subalpine fir, successional pine and aspen forests, and "man-made" grasslands/hay meadows. Large pure stands of lodgepole pine are found in the centre of the valley. Blocks of natural grassland are dotted throughout, on dry, warm, south-facing ridges. These natural grasslands are the most interesting features of the area. Kept in successional stages by both fire and now grazing, these spots support the highest diversity of any ecosystem in the area. Many characteristic species such as Rocky Mountain juniper are either rare or absent elsewhere in the region.

The grasslands, known locally as "Juniper Hills," are almost all privately owned. They are usually easy to reach but check with the owners before entering private property.

There is one spot, not privately owned, in the middle of the valley, which once supported a Forest Service lookout (Malkow Lookout). To reach this "bump," drive south of Smithers on Hwy 16 to just south of, or beyond, the Bulkley River bridge, and take the (old) Babine Lake Rd left (west) off Hwy 16 to McCabe Rd on your left. Head north and east on McCabe. Watch for a small side road leading off to the left (north) and up the hill to Malkow Lookout. Park at the sign blocking the road. This is private property. The owners do not mind hikers and naturalists passing through, but please respect their wishes as signposted. The top affords magnificent 360° views of the valley and surrounding mountains.

**Mammals**: black bear, mule deer and moose.
**Plants**: little penstemon, rough-fruited hairy bell, Rocky Mountain butterweed, creamy potentilla, nodding onion, several rock cresses, stonecrop, frog orchid, green rein orchid, needle grass plus several other grass species, several wormwood species, woodsia and moonwort ferns, spike moss and clematis.

## Babine Mountain Recreational Area

This spot is one of spectacular beauty and includes extensive areas of gently rolling subalpine and alpine meadows dotted with lakes and lush with flowers in late July and August. Many different birds are present in the high country.

Small parties of hikers are almost assured of spotting goats in summer or early fall on high scree slopes.

Gates are present on all roads at the boundary of the Recreation Area (approximately 1000 m) to restrict vehicle access. This is an attempt to reduce damage to the fragile alpine meadows and tundra. There are three access routes:

Driftwood Creek Rd can be driven to within 3 km of the boundary (about 6 km beyond Driftwood Creek Fossil Beds). To continue driving up the road to the boundary requires a 4-wheel drive vehicle and fording the creek.

The Cronin Mine Rd leaves Babine Lake Rd (Eckman Rd), 34 km from the junction with Hwy 16 east of Smithers. This Cronin Rd becomes very rough and requires a vehicle with good clearance. Drive for about 9.6 km to the mine buildings. A side road up Higgins Creek, 8.0 km from the turn off the Babine Lake Rd, provides easy hiking access to an old mine and the alpine area.

Mt Astlais, or Big Onion, an area set aside for snow mobiles, is gained from the Old Babine Lake Rd. Watch for signs along the road. The area is closed to vehicles in summer.

**Birds**: Golden- and white-crowned sparrows, water pipits, horned larks, ptarmigan, grouse, eagles and hawks.

**Mammals**: grizzly and black bear, coyote, mule deer, moose, porcupine and mountain goats.
**Plants**: paintbrush, fleabane, ragwort, larkspur, valerian, monkshood, false hellebore, columbine, cinquefoil, forget-me-not, rein (bog) orchid and whitebark pine (Cronin and Higgins Creeks).

## Telkwa Microwave Road (Winfield Creek)

This site provides easy car access to the high country. Once up there, hiking is relatively easy in the wide open gentle slopes of these alpine and subalpine areas.

The good all-weather road is maintained by B.C. Telephone, but you do not have to have permission to use it. The road takes off west from Hwy 16 at Telkwa village. Drive 26.5 km west to the turnoff to the north. Take the north road another 9.6 km to the parking area at the second microwave tower. This road, although rough, can be driven by two-wheeled vehicles, but not campers. The road beyond the towers is suitable only for a four-wheeled vehicle.

**Birds**: hawks, owls, ruffed, spruce and blue grouse, willow ptarmigan, white-tailed ptarmigan and rock ptarmigan.
**Mammals**: Grizzly, mountain goats, black bear and moose.
**Wildflowers**: paintbrush, ragwort, valerian, monkshood, fleabane, larkspur, mountain heathers and rein orchids.

## Hudson Bay Mountain

The massive spectacular mountain, with four peaks, dominates the valley. Two hanging glaciers are visible from Hwy 16 west of Smithers. The mountain is the easiest access to alpine and subalpine country in the region that borders Hwy 16. The gently rolling plateau, known locally as "The Prairie," provides an easy, short (two hour) hike to the alpine Crater Lake.

The subalpine meadows are reached at the far end of the parking area on the ski hill road.

Fossil beds are found east of the Lower Silvern Lakes which are accessible only by foot. Past mining activity has dotted the mountain with remnants of this work.

The main access is along the ski hill all-weather road west from Smithers. The road passes through the Smithers Community Forest, where there is a self-guiding nature trail. For Hudson Bay Mountain bear right at the forks (at 14 km) and travel another 9 km to ski hill parking lot. The left hand fork goes to the old Duthie Mine site where a trail takes off to Silvern Lakes behind the mountain.

**Birds**: ptarmigan, grouse, white-crowned and golden-crowned sparrows, horned larks, water pipits, finches, and great gray owls.
**Mammals**: hoary marmots, wolves, grizzlies, mountain goats, moose, porcupine, black bear and marten.
**Plants**: magenta paintbrush, forget-me-not, valerian, purple fleabane, rein orchid, inky gentian, heathers, alpine poppies, blue corydalis, sub arctic-alpine lichens and whitebark pine.

### Glacier Gulch and Twin Falls

The site offers a view of Kathlyn Glacier and Twin Falls coming off the ice. Cold moist air and drainage from this glacier create a local climate much wetter than the surroundings and hence produces vegetation more typically coastal than the Bulkley Valley. Birds are fairly numerous, with the calls of varied thrush quite common in spring. Black bear and moose are present but seldom seen. This wetter site produces western and mountain hemlock, salmon berry, and white moss heather, plants not common around here.

The trail up to the glacier is very steep and dangerous in places. It is recommended only for experienced hikers. To reach the site take the north fork at the point where Hwy 16 swings east around Kathlyn Lake, about 4 km north of Smithers. For the remaining 6.4 km drive north to just before the lake, swing west across the railway track and then north beside the railroad. Turn west on Glacier Gulch Rd and continue west to the Gulch and picnic site.

**Further Information**

There are several good reports on the vegetation of this area published by the Ministry of Forests. Map sheets include Smithers 93 L (1:250,000).

*Rosamund Pojar*

## KITIMAT

Located at the head of a long inlet, the Kitimat Arm of Douglas Channel, Kitimat is surrounded by mountains that reach beyond treeline to 1700 metres. The Kitimat River estuary provides a good food source for shore and water birds in spring and fall. A variety of orchids occur, and five different species of blueberries can be found.

Lying on the west coast of B.C. the inlet brings a variety of seabirds landward. The estuary holds shorebirds, fall and spring, as they pass to and from breeding sites. Virgin forests consist of huge cedar, hemlock, amabilis fir and Sitka spruce. A giant spruce, approximately 12 metres in circumference, has been set aside as a municipal heritage site. The tree can be viewed within walking distance of the city centre. Summers are too cool for Douglas-fir, even though it grows in the neighboring community of Kemano. The Alaskan blueberry is abundant on logged slopes from mid July on. Above timberline, unique bogs with yellow cedar and mountain hemlock can be searched out.

**Further Information**

Kitimat lies south of Prince Rupert at the end of Hwy 25. The Hirsch Creek Bridge Campsite lies immediately to the right of the bridge. Information is available at the Kitimat Centennial Museum (open 11 AM to 5 PM Tuesday to Saturday. Boat charters, from the Marina on Kitamat Village Rd, are available to see sights along the inlet.

*Kitimat Centennial Museum*

# AIYANSH (TSEAX OR NASS) VOLCANICS

The Aiyansh volcano and lava flow of 250 years ago is one of the most recent in Canada. Located north of Terrace, the site is one of the easiest recent volcanic eruption to reach in the province. Originating at a small cinder cone, lava flowed onto the Nass valley for 25 km downstream. Different types of lava are well exposed including "aa," the chunky lava and "pahoehoe," the pillow, ropey material.

The features of the lava flow are best seen from the air, at which elevation you can note how Lava Lake was dammed by the molton flow, and the Nass River pushed to one side of its valley. On the ground, the rock is sharp enough to cut boots.

Near the river, north of the "Y" in the road are tree wells and molds, remnants of the cottonwood forest which is only now coming back. From the road near Canyon City you can see pressure ridges, collapse depressions, and surficial wave patterns. The depth of the flow can be seen along its edge from the Canyon City foot bridge.

The most popular feature is the cinder cone where the eruption began. From this 100-metre-high cone on Crater Creek, lava flowed north down the Tseax River valley and out onto the Nass River flats 25 km downstream. For those with hard hats and flashlights, two large lava tubes (subsurficial conduits) may be explored immediately below.

**Further Information**

The lava beds are reached by travelling west of New Hazleton on Hwy 16 to the junction with Hwy 37. Turn north on 37 for 86 km to the Cranberry Junction. Take it south-west and look for the flows out on the Nass River flats. An alternative route is to drive north from Terrace, off Hwy 16, about 80 km on the gravel road to the site. These gravel roads are used by logging trucks. Watch carefully and keep your headlights on at all times. • There are no established trails to the cone. Beginning at the picnic area on Lava Lake, drive north about 1.5 km and take the logging road on the right past Ross Lake. This may be followed up the ridge above the creek, roughly paralleling it along the shoulder above. Keep right! The road ends in slash, but the cone may be seen upstream in the valley below. A steep angling descent on foot and a log stream crossing brings you to the cinder hill. • Further tourist information may be had by writing: Tourist Information, 4515 Keith Ave., Box 107, Terrace, B.C. or phoning (606) 635-2063.

# TWEEDSMUIR PROVINCIAL PARK

B.C.'s largest provincial park, Tweedsmuir is one of the least visited parks in the province and thus still offers a true wilderness experience. Over 300 trumpeter swans winter at Lonesome Lake. Douglas-fir dominates at lower levels with cedar/hemlock rain forests in mid levels which contrast with alpine meadows and glaciers.

The southern half of Tweedsmuir Provincial Park encompasses a range of ecosystems from coastal forest to alpine meadow, valley bottom to mountain peak. Approaching the park by road from the east, Heckman Pass has the lower elevations of the Engelmann Spruce/Subalpine Fir Zone. The road then descends quickly through dry lodgepole pine stands on Young Creek Hill to the fir and cedar forests in the Atnarko Valley.

The western portion of South Tweedsmuir lies in the Coastal Mountains which have rugged peaks with numerous glaciers and icefields. Hunlen Falls, at a 260 m single drop, is one of the highest in Canada, and lies in the upper Atnarko watershed of these ranges.

North Tweedsmuir includes part of the Kitimat Ranges which are usually steep sided with well-developed cirques on the northeastern sides. Small glaciers are common.

The eastern portion of North Tweedsmuir lies in the western extremity of the Nechako Plateau. The rounded summits of the Quanchus Ranges provide vast expanses of open alpine country, and adjacent valleys often cradle interconnected lakes and meadows.

**Birds**: bald eagles, pileated woodpeckers, varied thrush, sapsucker, golden-crowned sparrows and trumpeter swans (winter on Lonesome Lake in the Atnarko Valley), blue grouse, ospreys, waterfowl, willow ptarmigan, rosy finches.
**Mammals**: grizzly and black bears. mountain caribou (Rainbow Ranges), mountain goat, river otter, wolf, red fox, mule deer, marten, pika, moose, deer, woodland caribou, coyote, wolverine, mink, marten, skunk, hares and hoary marmot.
**Fish**: steelhead and cutthroat trout, chinook, sockeye, coho, chum and pink salmon, Dolly varden, rainbow trout, mountain whitefish, kokanee burbot, sucker, chub, and squawfish.
**Biogeoclimatic zones**: (south) Coastal Western Hemlock, Sub-Boreal Spruce, Engelmann Spruce/Subalpine Fir, and Alpine Tundra; (north) alpine tundra and sub-boreal spruce.
**Plants:** subalpine fir, lodgepole pine, trembling aspen, paper birch, lodgepole pine and white spruce, black spruce, douglas-fir, Western red-cedar, black cottonwood, scrub birch, willow, horsetail sedges, pondweed and pond lilies, giant sword fern, huckleberries, blueberries, raspberries, wild roses, thimbleberries, salmonberries, lilies, orchids, heathers, mosses and a wide range of wild flowers.

**Further Information**
An expanding network of trails is provided for hikers and horseback. Two of the most popular hikes are the Rainbow Range Trail and Hanlen Falls Trail. All of the park is foot accessible, with guided hikes available for those unaccustomed to wild places. • Tweedsmuir south is accessed from Hwy 20. The north part of the park is reached on secondary roads from Hwy 16 south from Vanderhoof, Houston and Burns Lake to the Oosta-Whitesail Lakes (Nechako) Reservoir; then by boat to the 350 m rail portage at Chikamin Bay. Float planes out of Burns Lake, Smithers, Vanderhoof and Bella Coola are another option. Campsites and boat launches are available, but no services are provided. • Tweedsmuir Lodge, and Tweedsmuir Wilderness Centre at Stuie, 12 km inside the western park boundary, are the main sources of accommodation. They each have a lodge, cabins and campgrounds. Maps include #92N, 93C-D-E-F, all 1:250,000.

*Dennis O. Kuch and the book*
The B.C. Parks Explorer *by Maggie Paquet*

# BELLA COOLA

Eulachon (*Thaleichthys pacificus*) is a small, smelt-like fish, found only on the west coast of North America. It has long been prized for its rich oily flesh. Eulachon and its oil are a staple in the diet of many coastal native people. Each spring the fish come in from the ocean to spawn in the lower reaches of large coastal rivers including the Bella Coola.

In prehistoric and historic times, the small fish eulachon was a major item of trade. Both dried fish and their oil processed into a butter-like grease moved inland via "grease" trails. The Nass River, north of Prince Rupert, is reported to have the largest run of these fish. The word "Nass" is a Tlingit word meaning "food depot." Trading trails to carry the grease inland and return trips with moose hides and other material not available on the coast, followed the Nass and Skeena River systems inland, creating one of the largest and most complex trail networks on the continent. The Bella Coola grease trail followed the Dean, Atnarko and Bella Coola Rivers as did Alexander Mackenzie with his Indian Guides in 1793.

The Thorsen Creek petroglyphs are a major grouping of rock carvings set on a high bluff above the cascading creek.

**Birds**: bald eagles, great blue herons, gulls, various sparrows, kinglets and warblers (from October to May). Trumpeter swans use the flats in winter.
**Marine Mammals**: seals.
**Mammals**: grizzly, black bears, mule deer, red foxes, wolves (Walker Island).
**Plants**: old-growth Western red-cedar (Walker Island Regional Park), sitka spruce, black cottonwood, devil's club, huckleberry, salmon berry, red alder, lupine, water parsnip and fritillaria.

**Further Information**
The community is accessed via Hwy 20, 480 km west of Williams Lake, and 53 km west of Tweedsmuir Provincial Park. Most services are available. The local natural history group, the Bella Coola Trail and Nature Club, may be contacted c/o Tony Karup, Bella Coola, B.C. V0T 1C0.

*Dennis Kuch*

# QUEEN CHARLOTTE ISLANDS

The Queen Charlotte archipelago has long been recognized as a great place for marine birds, marine mammals, fish and benthic communities. The lush forests, estuaries, sand dunes, wetlands, serene mountains and meandering rivers and streams all have abundant food sources to support this life. The bird checklist has grown to 240 species. The total breeding population of seabirds is estimated at over a million pairs. Bald eagles and peregrine falcons are numerous.

The archipelago consists of more than 150 islands, the tops of a submerged ridge, of various sizes some 50 to 130 km from the mainland. There are two major islands, Graham and Moresby, Graham to the north is the most accessible with long beaches on the east side. It has a rugged, rock-girded western shoreline. Moresby is more remote with high mountain ranges and spectacular old-growth forest. Some of the largest cedar and spruce trees in Canada can be seen here. Skidgate Channel slices between Graham and Moresby.

Moderating effects of the sea makes the climate quite different from the mainland. Though the climate of the "Charlottes" is mild, it has spectacular winter storms with hurricane force winds, wild seas and some of the highest rainfall in Canada — 450 cm.

Birds, particularly those of the sea and coast, are numerous. Incredible seabird colonies support numerous alcids that are unusual elsewhere in Canada. This is one of the few places in Canada where black-footed albatross can be seen just off Frederick Island on the northwest coast.

**Birds**: (species rare in Canada) red-faced cormorant,magnificent frigatebird, Steller's eider, bar-tailed godwit, Aleutian tern, horned puffin, yellow wagtail, red-throated pipit, great-tailed grackle, brambling; (others) four loon species, grebes, black-footed albatross, northern fulmar, shearwaters, Leach's and fork-tailed storm-petrels, double-crested, Brandt's and pelagic cormorants, tundra and trumpeter swans, five goose species including emperor, many ducks including Eurasian wigeon, common, king and Steller's eiders, bald eagle, northern goshawk, merlin and peregrine falcons, gyrfalcon, sandhill crane, black-bellied, Pacific golden, lesser golden, and semipalmated plovers, American black oystercatcher, wandering tattler, whimbrel, marbled godwit, ruddy and black turnstones, surfbird, red knot, sanderling least, Baird's, pectoral, sharp-tailed, rock, and buff-breasted sandpipers, dunlin, long and short-billed dowitchers, three phalarope species, jaegers, gulls, terns, alcid species including common and thick-billed murre, pigeon guillemot, marbled and ancient murrelets, Cassin's and rhinoceros auklets and tufted puffin, short-eared and snowy owls, and the Haida Gwaii endemic subspecies of northern saw-whet owl, hairy woodpecker, Steller's jay and pine grosbeak.
**Marine mammals**: Dall's porpoise, harbour porpoise, killer whale, gray and humpback whales, harbour seal, Steller's sea lion, and northern fur seal.
**Mammals**: black-tailed deer, beaver, muskrat, raccoon, red squirrel, Roosevelt elk, black bear, pine marten, Haida weasel, river otter, several mice, four species of bats, two races of shrews, and three rat species.
**Fish**: chinook, coho, pink, and chum salmon, rainbow, steelhead and cutthroat trout, Dolly Varden and stickleback.
**Amphibians:** Tree frogs (*Hyla regilla*), tree toad. No snakes, lizards or newts have made it to the Charlottes.
**Plants**: alpine daisy, sphagnum bog plants, mace-headed sage, American dunegrass, beach sage, sea rocket, beach glenia, beach tansy and Sitka spruce. West coast rain forest includes huge red-cedar, cypress, Sitka spruce and hemlock.

Vast dune communities are present along the beaches in Naikoon Provincial Park. These plants together form a stabilizing mat to hold the dunes.

Several insect species are endemic including a species of spittle bug and a dark carabid ground beetle. Only a few butterfly species are present, though there are some moths including a few endemics.

The eight-metre tidal range on the east coast creates incredible diversity in the intertidal area. Here you will find a very rich marine environment away from the freshwater influence of the mainland coast.

The geology, like that on the rest of the west coast, is complex, with plate tectonics and lava flows playing a major role. The basalt cliffs of Tow Hill jut skyward 109 metres revealing a remnant of volcanic activity. Seismic disturbances are common and measured at the station at the Queen Charlotte Islands Museum. The lowlands consist of glacial outwash plains with many small lakes and ponds.

## HECATE STRAIT: PRINCE RUPERT - SKIDEGATE CAR FERRY

The ferry from Prince Rupert is an excellent way to find numerous birds any time of the year.

**Birds**: (winter) common and Pacific loons, northern fulmar, short-tailed shearwater, Thayer's gull, black-legged kittiwake, common murre and pigeon guillemot, marbled and ancient murrelets, Cassin's and rhinoceros auklets.(spring - summer) four loon species, northern fulmar sooty, short-tailed, black-vented and pink-footed shearwaters, fork-tailed storm-Petrel, two cormorant species, three scoter species, red-necked and red phalaropes, parasitic and long-tailed jaegers, black-legged kittiwake and other gull species.

## MASSET

Nestled on a point where two major waterbird flyways (Masset Sound and Dixon Entrance) converge, the village of Masset offers many easily accessible birding opportunities. The seaplane spit, Masset Harbour, Delkatla Bay and Entry Point are all within walking distance.

## DELKATLA WILDLIFE SANCTUARY

This 288 ha of wetlands is one of the most important sites on the Charlottes for waterfowl and shorebirds. It is also a vital link in the chain of coastal marshes along the B.C. coast. Over 140 species of birds have been recorded in Delkatla since the 1890s.

The Sanctuary Committee has constructed a number of observation points including two towers and adjacent parking lots on Tow Hill Rd which runs east from Masset. Cemetery Rd runs north off Tow Hill Rd along the Sanctuary's eastern boundary. Trumpeter Drive runs along the southwest perimeter and affords excellent views of the main slough and extensive flats. The dike bisects the Sanctuary. A tower stands nearby. Check the freshwater slough on the northwest side of the Sanctuary for dabbling ducks and trumpeter swans in winter. Use the tower and viewing platform which are accessible from the dike road.

**Birds:** trumpeter swans, greater white-fronted geese, Canada geese, many duck species, bald eagle, peregrine falcon, sandhill crane, shorebirds including marbled godwit, semi palmated, western, least, and sharp-tailed sandpipers, ruff and both dowitchers. Most of the landbirds found on the Islands can be seen here.
**Plants**: grasses, herbs, rushes, *Juncus effusus*, sedges, alder, spruce, clover, sorrel, plantain and yarrow.

## ENTRY POINT

This point of land, located at the entrance to Masset Sound, is an excellent spot to watch seabird movements as the tide changes in the Sound and offshore in

Dixon Entrance. The fast-moving, churning waters of the Sound and nearby offshore waters are important feeding areas for many birds.

To reach this site, drive 3 km north from Masset along Masset Sound to the Haida Village. The pavement stops at the far end of the village. Turn left and park by the Old Masset Cemetery and walk the beach at low tide. This is Indian Reserve land. Please ask at the Band office for permission. A nearby slough often contains ducks and shorebirds. Check the spruce trees for land birds.

**Birds**: four loon species, grebes, cormorants, sea ducks, bald eagles, peregrine falcon, shorebirds, gulls, alcids, northern shrike and lapland longspur.

## Skonun Point

Skonun, Yakan and Tow Hill are the three important rocky points between Masset and Rose Spit for excellent viewing of sea and shorebirds along the northern coast.

**Birds**: common and red-throated loons, harlequin duck, three species of scoters, sandhill crane, semipalmated plover, black oystercatcher, black turnstone, sanderling and rock sandpiper.

Skonun lies approximately 4 km east of the Delkatla causeway on the Tow Hill road. Turn off to the left (north) on a dirt road just beyond the pavement and the Canadian Forces Operations site. Park on the grass before reaching the sand dunes.

## Naikoon Provincial Park

The park, set on the northeast side of Graham Island, lies on the Argonaut Plain formed from glacial deposits. The site provides a wide diversity of habitats. Today the plain is dotted with small lakes and dissected by meandering streams. Of the 110 km of beach within the park, Yukon Point and Tow Hill are the only rock formations providing both shelter and food for divers, shorebirds, cormorants and other waterbirds. Tow Hill, an outcrop of basalt columns rising 109 metres above sea level, is a prominent landmark along the northern boundary.

Upwelling currents, especially near Rose Spit, attract seabirds normally not seen near shore. Spring and fall migrations provide outstanding birding at Rose Spit and Tlell at the southeast corner of the park.

Sea mammals are best seen along the East Beach and Rose Spit. Vegetation in the park is Coastal Western Hemlock Zone that has been modified due to the extensive muskeg and bog areas. Vast dune communities occur along the beaches. Insects include endemic beetles found on the beach.

The park occupies part of the Hecate Depression, a trough between the Outer Mountains to the west and the Coast Ranges on the mainland to the east. Basalt rocks at Tow Hill were formed by lava which solidified into faceted basalt pillars about two million years ago. At the base of Tow Hill layers of even older lava flows are exposed.

Two ecological reserves have been established within the park: Yakan Point near Tow Hill for mossy spruce forests on old dunes; and Rose Spit with a huge meadow of dune flowers. In addition there are several other special areas including the Tlell River estuary with large sand dunes, a gently flowing tidal river, exposed seacoast and large areas of level upland. The road from Port Clements to Tlell, a former plank road, is very straight and provides easy access to sphagnum bogs with their special plants.

**Birds**: sandhill cranes, red-throated loons, semipalmated plovers and all Charlotte sub-species.

**Fish**: three species of salmon — coho, pink, and chum; steelhead, cutthroat trout, Dolly Varden, and stickleback.

**Plants**: stunted lodgepole pine, cypress, western red-cedar, western hemlock, Sitka spruce, salal, salmonberry, huckleberry, cranberry, red alder, Pacific crabapple, mace-headed sedge, American dunegrass, beach sage, sea rocket, and beach tansy.

## Rose Spit

To reach Rose Spit take the Tow Hill Rd, Hwy 16, east from Masset along the north coast for 26 km to Tow Hill. This gravel road passes through moss covered coastal forest, the Yakan Ecological Reserve, across the Sangan River estuary and terminates on the far side of the Hiellen River. Agate Beach Campground is found on the beach to the west of the perpendicular basalt columns. Continue on foot for the next 13 km along the vast sand beach to Rose Spit. Stay off the dunes with your vehicle. There are no facilities on the spit. Respect the fragile vegetation. Good maps and other information are obtained from park headquarters near Tlell on Hwy 16.

**Birds**: similar to other sites and excellent during migration.

## Langara Island

If you have time and the weather holds, a boat trip to Langara Island off the north coast of Graham Island is worth it. The island hosts a large seabird colony with thousands of nesting pairs of birds seldom seen near land except at their nesting colonies. On the way out you will encounter several rich seabird feeding grounds including the mouth of Masset Sound, Naden Harbour, Pillar Bay, Kiusta, Perry Passage, Cloak Bay and Cox Island. A trip off the west coast of Langara Island and then south toward Frederick Island provides an excellent chance of seeing black-footed albatross, fulmars and shearwaters. Plan at least four days for this trip and expect to see a wide variety of birds. Flesh-footed shearwater, horned and tufted puffin were seen here in 1988.

## Yakoun River Estuary

The Yakoun River is the largest and most important salmon and steelhead river on the Islands. It also supports the largest wintering concentration on the Charlottes. Extensive mud flats at its mouth, exposed at low tide, stretch for 2.5 to over 3 km covering Yakoun Bay in Masset Inlet.

The Golden Spruce, one of the most famous trees in B.C., grows along the west bank of the Yakoun River about 6 km south of Port Clements on the road to Juskatla. This tree, a Sitka spruce, has golden yellow needles where exposed to the sun. The start of the trail to the spruce is from a well marked pullout on the main road. The path winds through an unlogged forest of huge red-cedar, hemlock and spruce. At the end of the trail look across the brown river to see the 20-metre tall tree on the opposite bank.

A few km further south towards Juskatla is a Haida Canoe, the log it was shaped from, and the stump of the cedar, all of which clearly show the adze marks of the Haida canoe builders. A boardwalk and trail leads off the road to the site.

**Birds**: Brant, Canada geese, trumpeter swans, ducks and dunlin.

The Yakoun estuary is located on the south side of Masset Inlet on the southwestern perimeter of Port Clements. This logging community is on the main paved Hwy 16, 45 km south of Masset and 80 km north of Queen Charlotte Cit. Take either of the Port Clements turnoffs which eventually merge into the main street through the village. Easy access into the estuary is attained at the upper end.

## Skidegate Inlet

Skidegate Inlet and Skidegate Channel separate Graham and Moresby Islands. Winter birding is especially good as the inlet is somewhat protected from storms, and has several shallow, freshwater estuaries.

Gray whale watching is a spring pastime. They can be seen from the highway from April through June. Porpoises, seals and killer whales may appear anytime.

**Birds**: loons, grebes, brant ducks, bald eagles, black oystercatchers, gulls and pigeon guillemot.

## Skidegate Mission

This attractive Haida community lies at the northeastern entrance to Skidegate Inlet. Pigeon guillemot nest near here and can be seen from shore. The pebble beach is good for shorebirds. Large rafts of harlequin ducks and scoters are present in April and along the coast road north to Tlell. In winter western grebe and loons occur offshore. The Queen Charlotte Museum lies at the western edge of the Mission about one km east of the ferry docks.

## Queen Charlotte City

Most of the landbirds of the Islands can be found in or near this community. Rarities are a speciality here.

Tidal mud flats, including the Macmillan Bloedel dryland log sort near the western edge of town, are excellent places to look for Eurasian wigeon, other dabbling ducks, white-winged and other gulls and shorebirds.

Nearby Sleeping Beauty Mountain alpine areas have the endemic daisy, *Senecio newcombii*. To reach the meadows look for the turnoff on Crown Forest roads near the Honna River, east of Queen Charlotte City. Ask for directions in Queen Charlotte City at the Ministry of Forests' office. The routes are steep; however, the flowers are worth the trip in late summer.

Queen Charlotte is one of the larger centres on Graham Island. Kallahin Expeditions (Mary Morris) and other operators provide nature tours and expeditions from here.

**Birds:** (winter) black oystercatcher, hermit and varied thrushes, winter wren, dipper, Bohemian and cedar waxwings, orange-crowned, Townsend's and yellow-rumped warblers, fox song and golden-crowned sparrows and junco.

## Alliford Bay - Skidegate Landing Car Ferry

Crossing time is 20 minutes and offers great opportunities to see loons, grebes ducks and alcids. The ferry runs approximately hourly through the day and several times in the evening all year around.

**Birds**: loons, grebes, cormorants, ducks. pigeon guillemot and other alcids.

## Sandspit

Located on a point of land between Hecate Strait and Shingle Bay on Skidegate Inlet, the one km long spit juts east into the strait. Excellent birding is found around the airport. Look for seabirds including loons, grebes, ducks, shorebirds and alcids, all of which are common in winter. The largest concentration of wintering brant (over 300) in British Columbia frequents Shingle Bay. At low tide the vast expanse of pebble flats abound with shorebirds and waterfowl feeding at the tide line.

**Birds**: many waterfowl species including brant, sandhill crane, shorebirds, gulls and alcids; (winter) waterfowl, bald eagle, peregrine falcon, shorebirds including black-bellied plover, ruddy and black turnstones, surfbird, rock sandpiper and dunlin. This is a birding hotspot during migration.

## North Moresby Island

High mountain/alpine areas with good flowers in July and August are found south of Mosquito Lake on Mount Morseby. A salmon hatchery is at Pallant Creek.

## South Moresby

South Moresby is a collectio of islands, islets, and waterways, sheltered bays and inlets, freshwater streams and estuaries. There is an incredible variety and diversity of mountain, alpine, shoreline, island and marine environments. Outstanding cultural heritage sites and abandoned totem villages of the Haida are present. At Hotspring Island, you can bask in 38° C mineral hot springs.

Thousands of seabirds come to islands within South Moresby in summer to nest. Watch for young-of-the-year of rhinoceros and Cassin's auklet on the water in late summer. Horned and tufted puffins can be seen around

Flatrock Island and Adams Rocks and peregrine falcons are often seen around Kunghit Island. Watch for these falcons chasing eagles.

Dall porpoise will swim at top speed along side a boat in Juan Perez Sound, playing under the bow. Harbour seals haul-out on rocky shores and whales are often spotted.

For information about South Moresby/Gwaii Haanas National Park Reserve contact the Canadian Parks Service, Box 37, Queen Charlotte City, B.C. V0T 1S0, or phone (604) 559-8818.

**Further Information**
The islands are known for their extreme weather conditions at any time of the year. Aggressive Pacific storms, with winds and rain may hit at any time. Bring warm, waterproof clothing and footwear. Rubber boots are useful. • One way to reach the Charlottes and see the natural history including birds is aboard the B.C. Ferry from Prince Rupert. In summer there are six round trips per week and three trips during off-season. • Time Air runs daily Dash service between Vancouver and Sandspit. Waglisla Air offers daily service from Vancouver to Masset. North Coast Air Services and Trans-Provincial Airlines provide several daily flights between Prince Rupert and Masset, Queen Charlotte City and Sandspit. • Many tour operators run birding and other natural history tours including Kallahin Expeditions at Box 131F, Queen Charlotte City, B.C. V0T 1S0. Phone 559-8455 or 559-4746. Pelagic and land birding trips can be arranged through Margo Hearne at Delkatta Bay Birding Tours, Phone 626-5015. Haida Gwaii Watchmen have guides stationed at several of the old village sites around the Islands. Islands Ecological Trips provide natural history tours island wide. Write Box 970, Queen Charlotte City, B.C., V0T 1S0. Phone 559-8648. • The Queen Charlotte Islands Museum, located on the Second Beach at Skidegate, about one km east of the Skidegate ferry dock, has an excellent Haida Gallery and natural history section featuring many of the plants and animals found on the islands. Numerous books have been written on the Queen Charlotte Islands including a detailed flora. General travel information is available from the Q.C.I. Chamber of Commerce, Box 38, Masset. B.C. V0T 1M0.

*Mary Morris, Margo Hearne and Peter Hamel*

# NORTHERN BRITISH COLUMBIA

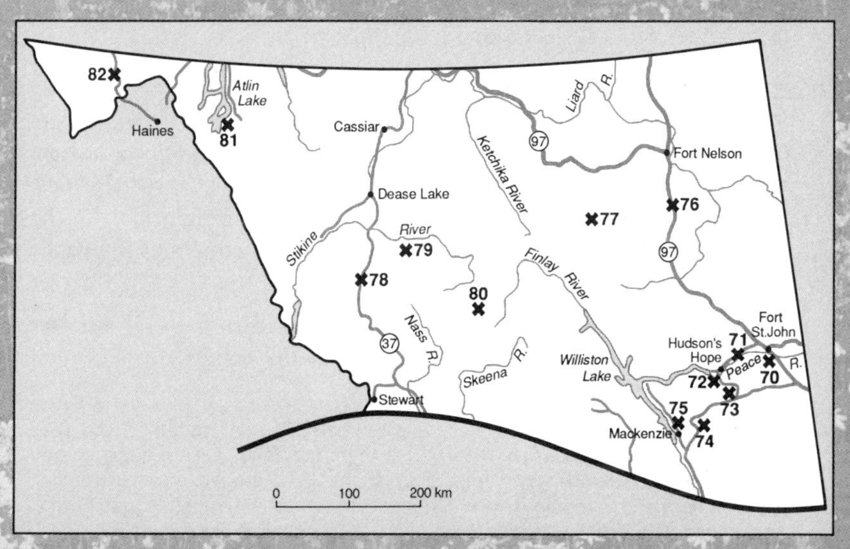


70. Fort St.John
71. Canoe Trips Hudson's Hope to Fort St.John
72. Hudson's Hope
73. Moberly Lake P.P.
74. Pine Pass
75. Mackenzie/Southeast Willston Lake
76. Alaska Highway North of Fort St.John
77. Kwadacha Wilderness P.P.
78. Stewart - Cassiar Hwy 37
79. Spatsizi Plateau Wilderness P.P.
80. Tatlatui P.P.
81. Atlin P.P.
82. Haines Road

# FORT ST. JOHN

This area, located at the junction of aspen forest/grassland and northern forests, attracts eastern birds such as blue jays and grassland birds such as sharp-tailed grouse.

## STODDARD ("FISH") CREEK

This site is an extensive white spruce and aspen stand. Drive north along 100 St through Fort St John to the B.C. Rail tracks (don't cross them). Turn east, or right, onto 146 Rd and drive 0.5 km east. Turn left to cross the tracks and park in the College parking lot.

## ST. JOHN CREEK AND THE NORTH PINE AREA

Proceed north along 100 St, over the tracks and Stoddard Creek bridge. This is 101 Rd, a rural continuation of 100 St running north for at least 60 km to Prespatou. About 6 km north of Stoddard Creek, the road divides. Take the right branch descending into the valley of St John Creek. The woods along the creek at the bridge are a good spot for migrants from mid-August.

Back on 101 Rd, drive up the hill, and turn right (east) onto 112 Rd, and drive 3.5 km east, and turn left (north) onto 259 Rd. The next several kilometres are good for birds.

**Birds**: blue jay, Philadelphia vireo, black-throated green warbler, Cape May warbler, clay-colored sparrow, redpolls, northern goshawk, golden eagle, snowy owl, northern hawk-owl, pine grosbeak, sharp-tailed grouse, northern shrike, pintail, tundra and trumpeter swans, Canada geese, harriers, rough-legged hawks, golden and bald eagle, occasional peregrine falcon, Lapland longspurs.

## SOUTH SEWAGE LAGOONS

The sewage settling ponds lie about 0.5 km south of the Alaska Hwy , near the south edge of town. They attract waterfowl, raptors and hordes of migrating song birds. In late May songbirds can fill the weedy margins. Blackpoll and Tennessee warblers can be common, with surprises like golden-crowned sparrow or an occasional Harris' sparrow in fall.

**Birds**: tundra swans, Eurasian wigeon, hooded merganser, blue-winged teal, northern shoveler, canvasback, lesser scaup, common and Barrow's goldeneyes, ruddy duck, white-rumped, pectoral and Baird's sandpipers, lesser golden and black-bellied plovers, lesser yellowlegs, long-billed dowitcher, semipalmated and least sandpipers, red-necked phalarope and Wilson's phalarope.

To reach the lagoons, drive south along the Alaska Hwy past the Totem Mall (on your left) and look for a gravel road to the right within a km of the mall. This road is marked with a large sign showing two dirt bike racers. Follow this road as it turns right. At the bus company building turn left and follow the road past a ball field up to the dikes of the lagoon.

## North Sewage Lagoons

These five ponds are surrounded by fields, lying northeast of Fort St John, and northwest of the airport. All birds mentioned for the South Lagoons (see above) are present, but usually in greater numbers and in more variety.

This is a good spot for rare birds like brant, wood duck, greater scaup, and Sabine's gull.

**Birds**: sharp-tailed grouse, northern shrike, several species of sparrow, hudsonian godwits, buff-breasted sandpiper, whimbrel, white-rumped, stilt sandpipers and ruff.

Drive east out of Fort St John toward the airport along 100 Ave. Part way through the "S" curve west of the airport, look for road 259 to the north (left). Take this road for about 1.5 km north to the lagoon/dikes.

## Attachie Lookout and Upper Cache Road

Lying 40 km west of Fort St John, along Hwy 29, the site looks out over the Peace River. Fields below Attachie Lookout attract flocks of Canada geese each spring. Mule deer are numerous on the breaks below the lookout. In spring watch for pasque flowers (crocus).

**Birds**: bald eagle's, Vesper and Clay-colored sparrows, orange-crowned warblers, northern waterthrush, swamp sparrow, common grackle, hermit thrush and the odd veery, kestrel, white-winged crossbill, hawk-owl, goshawk and golden eagle.
**Mammals**: Moose, mule deer and coyote.

The Upper Cache Rd (116th Rd) branches north off Hwy 29, about 100 metres east of Attachie Lookout. This 30 km gravel road passes through good birding areas.

## Beatton River Provincial Park

One of the best birding spots from May to October, the small spruce grove attracts a few eastern warblers. Small flocks of migrant songbirds pass through in May and August through early September.

During wet summers a variety of mushrooms and other fungi appear in the spruce grove.

**Birds**: ruffed grouse, Cape May warblers, bay-breasted warbler, black-throated green warblers, boreal chickadee, gray jay, golden-crowned kinglet, olive-sided flycatcher, Townsend's warbler, black-backed woodpecker, blackpoll warbler, magnolia warbler, Canada warbler, Philadelphia vireo, white-winged and surf scoters, horned, eared, and red-necked grebes, common, Pacific, and red-throated loons, oldsquaw and both goldeneyes.
**Mammals**: beaver, pine marten.
**Fish**: walleye and northern pike.

The park is found on the eastern side of Charlie Lake, 13 km north of Fort St John. Drive north from Fort St John along the Alaska Hwy for about 5 km to paved 271 Rd. Follow the signs for about 8 km to the park.

## Charlie Lake Provincial Park

Look for ruffed grouse, and warblers in migration. Many water birds can be seen from the boat launch. At least 10 species of gulls pass through here regularly including a few Sabine's gulls every year since 1982. Parasitic jaegers have occurred a few times.

Forest cover is aspen, with stands of birch, lodgepole pine and spruce.

Located 10 km northwest of Fort St John, the park is reached from the Alaska Hwy.

## Peace Island Park Road and Regional Park

Peace Island Park Rd lies on the south side of the Peace River and passes through mixed woods where eastern birds are more readily found than elsewhere around Fort St John.

During the height of migration, May 20 to June 1, and August 18 to 31, rarities to look for include Cooper's hawk, calliope hummingbird that nested here in 1987, white-breasted nuthatch, gray-cheeked thrush, Philadelphia vireo, Cape May warbler, ovenbird, MacGillivray's warbler, and golden-crowned sparrow.

**Birds**: Three-toed woodpeckers, while downies, hairies and pileated woodpeckers, Eastern phoebes, Tennessee, orange-crowned, magnolia, black-throated green and Canada warblers, great-horned owls, white-winged crossbill, common and hoary redpolls, Canada geese, American kestrels, ruffed grouse, barred owls, northern saw-whet owls, alder and least flycatchers, blue jays, black-and-white and mourning warblers, and other song birds.
**Mammals**: Moose, black bears, short-tailed weasel, red squirrel and bats (*Myotis* sp.)

Peace Island Regional Park is divided into a lawned picnic site, camping area, and a large riparian grove.

The park lies 12 km south of Fort St John, and 0.8 km west of the Alaska Hwy. Drive south from Fort St John on the Alaska Hwy and take the first turn-off at the southern end of the Peace River bridge.

## Kiskatinaw Provincial Park

This site, 15 km south of Taylor, has eastern birds nesting in its white spruce grove. Reach the park by a 6-km paved section of old road about 25 km north of Dawson Creek.

## Swan Lake

This site, on the B.C. border 38 km south of Dawson Creek, is noted for having Leconte's and sharp-tailed sparrows found in the large meadow bordering the lake.

To reach the site, take Hwy 2, 38 km south of Dawson Creek to the narrow road marked 201 on the west side of the border. This 201 road crosses the railway tracks and takes you to the lake, and then turns east. Park at the east turn.

Walk along the cutline through the grove of trees. Look to the left for the meadows. Cross the fenceline and enter these meadows to find the sparrows.

### Cecil Lake and Boundary Lake

Both lakes lie east of Fort St John; Cecil about 20 km, and Boundary on the Alberta border about 70 km east of town. Cecil Lake is best explored by canoe. Boundary has three oil lease roads that run on dykes into the lake. The flood plain of the Beatton river has two good birding sites: the arid steep hills north of the Regional Park and the cool mixed aspen woods on the south side of the river.

**Birds**: red-necked, horned, western and eared grebes, Trumpeter swans, tundra swans, dabbling and diving ducks, bald eagles, northern goshawks, broad-winged hawks, migrant shorebirds, LeConte's and sharp-tailed sparrows, rusty blackbirds, and palm and blackpoll warblers.
**Mammals**: muskrats, mink, porcupine and coyotes.

To reach the lakes, drive east from the centre of town along 100 Ave. This street becomes 103 Rd and runs by the airport. With the airport property on the south, you come to a T-junction; turn left or north and soon this road goes into the Beatton River canyon. Continue across the bridge on the paved road, and climb out of the canyon to the tiny community of Cecil Lake. Then turn around and retrace your route back west for about 0.3 km to the one lane dirt road that branches north off 103 Rd., for about 1 km to the southern end of Cecil Lake. If it looks like rain, or the dirt road is damp, stay off of it and walk the 1 km to the lake.

For Boundary Lake, continue through the community of Cecil Lake for another 30-40 km, through Goodlow, until the paved road turns south in a bend. Halfway through this bend, turn left (east) onto a gravel road. Follow the gravel road east past a small airfield until you reach a stop sign; turn left (north) and drive less than a km to a road to the east. This road has a Texas gate where it joins the gravel road. This is the southern of the three oil lease roads that run as dikes out into the lake.

*Chris Siddle*

## CANOE TRIP ON THE PEACE RIVER —HUDSON'S HOPE TO FORT ST. JOHN

This is an easy 80-km trip. Launch from Hudson Hope, below the ledge just east of Alwin Holland Park. The trip can be done in a long day, but two days is better. The valley acts as an east-west migration corridor for several species of waterfowl, and some of the best birding in mixed forests in the region are found here, especially for such "eastern" species as Canada warbler. Large mammals cross the wide gravel plains.

Floating down the river, you can notice the difference in vegetation between the shaded, cooler south side of the river, and the sunnier, drier north side. Pockets of natural prairie, locally known as "breaks," cover these south facing slopes. They are fascinating micro climatic sites where avens, nodding onions, cacti, roses and bedstraw grow.

**Birds**: Tundra and trumpeter swans, Canada geese, pintail, surf and white-winged scoters, herring, ring-billed, and Bonaparte's gulls, common and Arctic terns, eastern phoebe, orange-crowned warbler, fox sparrow, black-throated green warbler, least flycatcher, yellow-bellied sapsucker, rose-breasted grosbeak, ovenbird, common nighthawk, mourning warbler, red-eyed vireo, western flycatcher, western tanager, magnolia warbler, black-and-white warbler, Lincoln's sparrow, glaucous, Thayer's and Sabine's gulls, Arctic tern, red-breasted merganser, oldsquaw, common merganser, common goldeneye, mallard, ruffed grouse, common and hairy redpolls and pine grosbeak.
**Mammals**: moose, mule and white-tailed deer, elk, black bear, coyote, fox, lynx, pine marten, short-tailed weasel, and woodchuck.
**Amphibians**: Western toad and northern wood frog.

**Further Information**
Canoes may be launched from Hudson Hope, Lynx Creek, or Farrel Creek. The trip can be terminated at the mouth of Halfway River (42 km west of the Alaska Hwy), at the Old Fort (which requires scouting prior to the trip), or at Taylor Landing Provincial Park at Taylor south of Fort St John.

*Chris Siddle*

# HUDSON'S HOPE (ALWIN HOLLAND REGIONAL PARK)

This small park, about 3 km upstream from Hudson Hope, is one of the very few places where one can easily walk to the banks of the Peace River.

Canada geese nest offshore on the "teapot" rocks and are very conspicuous. A few violet-green swallows, uncommon this far east, can be found along with cliff swallows.

Young aspen trees form a fairly uniform forest east of the campground with a few birches and willows along the damp courses.

**Birds**: eastern phoebes, America dipper, common merganser.
**Mammals**: mule deer and bushy-tailed wood rats.
**Reptiles**: garter snakes.

**Further Information**
The park is reached by a short gravel road just off Hwy 29 between Alexander Mackenzie Bridge and Hudson Hope. Camping is permitted in a designated area within the park.

*Chris Siddle*

# MOBERLY LAKE PROVINCIAL PARK

Set on the south shore of Moberly Lake in the Moberly River valley adjacent to the Rocky Mountains, the park is one of the best spots in northeastern B.C. to find breeding black-throated green warbler. The rich variety of songbirds is enhanced by the mix of "eastern" and "western" species.

**Birds**: least flycatcher, white-throated sparrow, rose-breasted grosbeak; three-toed woodpecker, boreal chickadee, varied thrush, Hammond's flycatcher, bay-breasted warbler, MacGillivary's warbler common, and mourning warbler.
**Fish**: Dolly Varden, lake char, and lake white.
**Mammals**: moose, black bear, red squirrel, snowshoe hare and beaver.
**Amphibians**: long-toed salamander, western toad, spotted frog, wood frog, common garter snakes and western terrestrial garter snake.
**Plants**: white spruce, aspen, cottonwoods, red-osier dogwood, sarsaparilla, prickly rose, black twinberry, currant, high-bush cranberry, twinflower and dwarf blackberry.

**Further Information**
The park lies 3 km off Hwy 29, 24 km northwest of Chetwynd.

*Chris Siddle*

# PINE PASS

This region, extending for about 140 km along the John Hart Hwy 97 from the south end of Williston Lake (Parsnip River) north to Chetwynd, contains a low pass through the northern Rocky Mountains. The area contains a wide diversity of habitats. Several eastern and western plant and bird species meet along here, and a few "southern" species of birds probably reach their northern breeding limits.

A variety of boreal species occur year-round. Monkshood and chocolate lily can be seen just off the highway at the roadside stop above Azouzetta Lake.

**Birds**: three-toed woodpecker, spruce grouse, northern hawk-owl, common and hoary redpolls, Rosy finches, Canada warbler, mourning warbler, bay-breasted warbler, ovenbird, western flycatcher, Steller's jay and American dipper.
**Mammals**: moose, black bears, grizzly bears and Melanistic woodchucks.

**Further Information**
The John Hart Hwy 97 runs along the valley floor. Access to the alpine country involves a short but tough hike. The easiest way up appears to be the mountain slopes south of the highway just east of Azouzetta Lake.

*Chris Siddle*

# MACKENZIE / SOUTHEAST WILLSTON LAKE

Excellent views can be had along the Rocky Mountain Trench including much of the east side of Williston Lake. A variety of waterfowl and shorebirds come through in fall and spring.

This is also the northern limit of Douglas-fir. Plants range from bog communities to alpine meadows lush with flowers.

This is part of the Trench that was used by Alexander Mackenzie and John Finlay on their explorations of the west.

**Mammals**: moose, mountain caribou, grizzly, beaver, otter, mink, marmot, pika, Rocky Mountain bighorn and stone sheep.

## GATAIGA LAKE RECREATION SITE

The area is teaming with marsh-loving waterbirds and mammals. Plants include those most common to a bog type community such as swamp laurel.

**Birds**: mallard, green-winged teal, bufflehead, Barrow's goldeneye, bittern, snipe, Canada geese in fall, northern harrier (marsh hawk) and Whistling swans.
**Mammals**: beaver, muskrat, otter and mink, and moose.
**Fish**: rainbow trout and Rocky Mountain whitefish.

This day-use site has parking and a place to off load boats or canoes. The lake lies about 500 m off Hwy 39, on a short gravel road. The well marked turnoff lies about 16 km south of Mackenzie

## MORFEE MOUNTAIN

The mountain overlooks Williston Lake, the largest dammed lake in North America. On top you can view as far north as Peace Arm and as far south as the Parsnip River area. The mountain view from the edge of the Rocky Mountains allows a visitor to appreciate the immenseness of the Rocky Mountain Trench.

Surrounding Morfee Mountain is a complex of interconnecting ridges stretching for kilometres. Small ponds, created by snow melt, attract wildlife.

**Birds**: Stellar and gray jays and ptarmigan.
**Mammals**: marmot, pika, grizzly, moose, wolverine, mountain caribou, Rocky Mountain bighorn sheep and stone sheep.
**Plants**: heather, lupine, dwarf blueberry, alpine meadow species and Douglas-fir.

The mountain, accessible to the public, is owned by the crown, B.C. Hydro and B.C. Tel. Both companies have towers on top. To reach the site, drive 1.5 km north of Mackenzie on the only road out of town. Turn right at the B.C. Hydro and B.C. Telephone storage sheds and go up a rough winding road for about 16 km to the two microwave towers. This road is accessible by a two-wheel drive.

### Heather-Dina Lakes

This is the northenmost of the three sites. It offers one of the only full circle canoe routs in the Prince George region. The approximately 12 km route has few portages with the longest 500 m. Stopover sites are available for day and overnight along the route. The route is teaming with waterbirds from spring to fall. The site covers some 18 square km of lakes, creeks and portages.

**Mammals**: beaver, muskrat, otter, moose, black bear and Mountain caribou.
**Fish**: Rainbow trout.

**Further Information**

There are two launch spots, one at Heather Lake at 25 km on the Finaly Forks logging road, and the other at Dina Lake at 31 km on the same road. This logging road is winding and narrow in spots, but can be driven by car using caution at corners. The road begins immediately north of Mackenzie, and is in fact an extension of Hwy 39 from the south. • The topographic map for the area is Mackenzie, sheet 93-0, scale 1:250,000.

*John Crooks with thanks to Jim Tuck*

# ALASKA HWY NORTH OF FORT ST. JOHN

Much literature on this highway is available from motor clubs and other sources. We have selected a few sites that are of particular interest to a traveling naturalist.

### Pink Mountain

Natural history of this area is poorly known; however, from what is known this isolated outlier of the Rocky Mountains should be explored more thoroughly.

July is the time to see the mountain top covered with alpine wildflowers.

**Birds**: Rock, willow and white-tailed ptarmigan have been reported. Ruffed grouse, blue grouse, horned larks, water pipits, Bohemian waxwings, Wilson's warblers, and golden crowned sparrows.
**Mammals**: caribou, moose, black bear and hoary marmots.

To find the spot drive to the tiny community of Pink Mountain at Mile 143 of the Alaska Hwy, and then another 4 km to the Beatton River bridge at Mile 147. The road to the mountain takes off to the west from the bridge and is a 15-km drive. There is a fire tower on top, so the mountain is usually, though not always, accessible by four-wheel drive in summer. Ask directions, and update of road conditions from one of the several local gas stations.

## Clarke Lake Area

This region consists of black spruce and muskeg that is crossed by gas- and oil-lease roads. It is a good birding location so watch for nesting sandhill cranes in summer.

**Birds**: solitary sandpipers, lesser yellowlegs, common snipe, rusty blackbirds, common grackles, fox sparrows, Bohemian waxwings and Tennessee warblers.

About 8 km south of Fort Nelson, at Mile 295 of the Hwy, turn east at the Husky service station and drive down the Mile 295 Rd to cross the Fort Nelson River. Once on the other side, explore both the left and right forks of the road, but don't block traffic. If the road is wet, four-wheel drive is recommended.

## Fort Nelson and Airport

The site lies in the Fort Nelson Lowlands of heavily forested land. The area is relatively rich in eastern songbirds.

**Birds**: black-and-white warblers, white-throated sparrows, Philadelphia vireo, Canada warbler, Cape May warbler and mourning warbler, three-toed woodpecker and boreal owls.

The airport, about 5 km north of town, is one of the easiest spots to find these and other species. Follow signs from town, and once on airport property, park by the employee housing. Walk south past a strange white-domed building.

## Parker Lake

The shallow boreal lake is surrounded by muskeg. The muskeg contains a few yellow-bellied flycatchers. This is the only well-known and easily reached population of these birds in the province. Listen for the yellow-bellies about halfway down the narrow road and watch for them on the top of snags. Other birds to look for include palm, blackpoll and magnolia warblers.

To find the trail, set your odometer to zero in the centre of Fort Nelson and drive 11.2 km west and watch for the dirt road on the left at Mile 307. The lake is 1 km south of the highway, 7 miles west of Fort Nelson.

## Kledo Creek

The thick spruce woods found south of the roadside pulloff are very typical of boreal forest and host numerous eastern birds.

**Birds**: Cape May and bay-breasted warblers, varied thrush, greater yellowlegs, dusky flycatchers and rose-breasted grosbeaks.

To find the spot drive to Mile 335, 50 km west of Fort Nelson and watch for the small camping area beside the Kledo Creek bridge. At this location a road leads to a series of gravel pits about 1 km south of the highway. The spruce woods are near the gravel pits. Watch for gravel trucks.

*Chris Siddle*

# KWADACHA WILDERNESS PROVINCIAL PARK

Kwadacha is near the northern end of the Rocky Mountains immediately east of the Rocky Mountain Trench. The park contains some of the peaks and icefields of the Muskwa Ranges, and also straddles the divide between the Kwadacha and Muskwa Rivers.

Ninety-two species of birds and 15 mammals have been reported from the park. Three biogeoclimatic zones are found in the park. Patches of Boreal White Spruce and Black Spruce zone are at low elevation with virgin stands of spruce and fir. Above that is the Spruce/Willow/Birch zone. Alpine tundra tops the mountains.

Lloyd George Icefield, the largest north of 54° in the Rocky Mountains, feeds most of the rivers including the Kwadacha and Muskwa.

**Birds**: red-throated and Pacific loons, black scoter, peregrine falcon and numerous tree sparrows, common, Pacific and red-throated loons, red-necked grebe, several ducks including black scoter, merlin, golden eagle, willow and white-tailed ptarmigan, red-necked phalarope, Wilson's warbler and tree sparrows.
**Mammals**: moose, elk, caribou, goats, grizzly bear, wolf, wolverine, marten and short-tailed weasel.
**Amphibians**: northern toad (*Bufo boreus*).

**Further Information**
The closest road or centre is Fort Nelson, about 155 km to the north of the park. Access is usually by float plane as there are no roads near or into the park. Horseback or hiking in can be arranged through a professional guide service. For such services contact the Peace-Liard District office at Charlie Lake Park, SS #2, Comp. 39, Site 12, Fort St. John, B.C. V1J 4M7 or phone them at (604) 787-3407. • Topographic map is 94F 10 • Provincial Museum reports by Cooper *et al* and the book *The B.C. Parks Explorer* by Maggie Paquett. • Stone Mountain Provincial Park, Muncho Lake Provincial Park, Wokkrash Recreation Area, and Liard River Hot Springs Provincial Park are all sites on or near the Alaska Hwy and North of Fort Nelson that are worth stopping for. Details for each are available from B.C. Parks District Office at 9512-100 Street, Ft. St. John, B.C., V1J 3X6, Phone (604) 787-3407. Maggie Paquett's book *The B.C. Explorer* is another good source for these sites.

# STEWART—CASSIAR —HWY 37

Northwestern B.C. is traversed by only one transportation link, Hwy 37, which cuts through several river valley systems and is mostly paved. At Dease Lake you leave the Pacific watershed and enter the Arctic drainage system ending in the MacKenzie River.

Special sites include Bear Glacier, 20 km west of the Meziadin junction on the road to Stewart, and Mount Edziza, a volcanic mountain range around Km 440 and Spatsizi Provincial Park.

Before heading north make sure to have two spare tires and other emergency equipment since garages are few and far between.

**Km 0.0:** On Hwy 16 Kitwanga (pronounced KIT wan GA) is the first community. Turn right off highway immediately after crossing the Skeena bridge to see totem poles, including the pole of the Captured Medeek. This mythical monster, in the shape of a grizzly bear pierced by the top of the totem pole, is easily recognized. The main street goes through the village and rejoins the highway. To see the Indian fort "Battle Hill" archaeological site, leave the highway again, on the left, passing Kitwanga General Store and Esso station, and pull out at the hand-of-history marker. Here Indian knights in armour, complete with great copper shields, defended an earth-and-wood castle where women and children fled through trap doors to the safety of underground passages, storerooms, and hidden exits. The armour was made from hardwood, brine-toughened doubled caribou leather and copper. The Greasetrail, which the fort defended and which the modern highway follows, was the great "oil shipping route" of the past. Oolichan fish oil was packed to the interior over these trails in 1x1x1 m bentwood boxes. The road rejoins the highway.

**Km 38.2:** Kitwanga Lake. Mountain goats are sometimes on the cliffs across the lake. Vegetation is typical of wet interior cedar-hemlock forest.

**Plants:** salmonberry, baneberry, twisted stalk and bunchberry.

**Km 155.0:** Nass River bridge rest stop.

Meziadin Fishway: The outlet of Meziadin Lake is the Meziadin River, which drops over a natural cataract and almost immediately joins the Nass River. This cataract, more a waterfall than a rapid, was so difficult for salmon to ascend that relatively few were successful before construction of the fishway (fish ladder). The small patch of flat land at the base of the falls was a traditional meeting place of Tahltans from Telegraph Creek and Nisga'as from the Nass, to trade and hunt. To reach the fishway, take the first left north of Nass River bridge. Turn left again right away where the logging road crosses an airstrip and drive to the end of the airstrip. Park before the road goes down a steep grade, to be sure it is passable. The fishway is at the bottom of the hill.

Each fall salmon ascend the fishway and spawn in tributaries of Meziadin Lake. Hundreds can be seen from the first Hanna

Creek bridge, on the highway, 10 km north of the fisheries turnoff at the Nass bridge and at Tuitina Creek. These salmon spawning creeks around Meziadin Lake are all good grizzly feeding sites and should be treated with care.

**Km 163.0:** Access to Meziadin Lake Provincial Park. Moose and black bear are prevalent.
**Birds**: Bald eagles, water birds, warblers and sparrows.
**Plants:** Cedar-hemlock forest, rose, twisted stalk, false Solomon's seal and false azalea.

**Km 169.0:** Meziadin Junction and turnoff to Stewart, B.C. and Hyder, Alaska. Gas food and lodging are available here. There is a provincial campground nearby at Meziadin Lake Provincial Park. Stewart is 67 km to the west. Bear Glacier is about 20 km west of this junction.
The side trip to Stewart could be the highlight of the trip. The black shales and grey waters of the Nass drainage give way to the granite batholith of the Coast range, with sparkling clear streams falling from the icefields above. Watch for flocks of black swifts along Hwy 37A toward Bear Pass.
The famous Bear Glacier is the first and largest of several ice tongues of the Cambria Icefield. Twenty-three km from Mezidian junction, a short paved access road leaves the highway to the left, heading to a beautiful overlook and picnic site perched on a terminal moraine. The Cambria Icefield is larger than the Columbia Icefield of Jasper/Banff.

**Km 25.0** On the Stewart Rd: A roadside pulloff directly in front of the ice. The present site of the highway was covered by ice in the 1950's.
Along the Bear River toward Stewart bald eagles may be seen in Coho salmon season, usually late August to mid September. Fish Creek, near Hyder, is an even better place for eagle-watching.
Stewart is on Portland Canal, an arm of the Pacific Ocean. Two good motels, a hotel and other services are found. Begin your visit at the museum which is also a tourist information centre. Phone number is 636-2568.
The salt marsh at Stewart is a good place to find birds especially during fall and spring migration.
Superb hiking paths, following old mining trails built for horses in the early days, lead to some of the most dramatic scenery Canada has to offer. The Ministry of Forests and the Stewart Lions Club have brushed out and posted several trails and prepared a brochure which is available at the museum.
At Stewart the Municipal District provides a campground with sani-station.

Now return to Hwy 37. North from Meziadin Junction the next 200 to 250 km is the wildest and least populated area of the trip. Hemlock trees give way to huge subalpine firs, *Abies lasiocarpa*, which the loggers call "Balsam." In September bright red high-bush cranberries are everywhere in the Bell-Irving Valley. Lush dense undergrowth of devil's club, red elderberry and alder indicate the dampness of the climate. Rain can be expected all summer, and heavy snow in winter. Porcupines, groundhogs, and several kinds of mice and lemmings are seen crossing the road, especially when traffic is light. In late winter, when the snow is deep, moose will gather on the plowed roadway where they have a better chance of outrunning wolves. Travellers are not likely to see wolves, although they are plentiful. Their eerie calls may be heard on moonlit nights in winter and at dawn in early spring, and their tracks can be found wherever there is mud or fresh snow.

**Km 205.0:** Bell-Irving River crossing #I has a level paved rest stop.

**Km 262.0:** Hodder Lake is a designated rest area with tables, pit toilets and boat launch.

**Km 264.0:** Bell-Irving River crossing #2 has a telephone, food, gas and lodging.

**Km 276.0:** Ningunsaw River Valley. Mammals include moose, black and grizzly bear. Scan the alpine country for stone sheep and goats. The valley vegetation is in transition from the cedar-hemlock forest through which you have been travelling into northern spruce-willow-birch forest. Understory plants include false Solomon's seal, saxifrage and baneberry. Alder-covered avalanche tracks occur along the road and up the sides of the mountain.

**Km 300.0:** A major avalanche swept across the highway, covering it in 30 metres of debris in January 1989.

The geological forces become more and more apparent through the Ningunsaw. Horizontal beds give way to steeply dipping layers and fold patterns visible in road cuts and on the mountains above timberline. These patterns are especially noted in early fall when they are outlined in light snow. Steeply dipping beds of mudstones and conglomerates are bent into almost vertical bands above Echo Lake, a geologic pattern that marks the wide Iskut valley north of Bob Quinn with parallel ridges and swamps, like corduroy.

**Km 304.0:** At Echo Lake, a roadside historical marker plaque commemorates the Yukon Telegraph Line. Two line cabins may be seen partly submerged in the lake. Margaret Vanderberg

**Km 310.0:** Site of the former Bob Quinn Lake Hwy Maintenance Establishment, closed in 1989. The busy airstrip nearby, serves gold mines to the west. The road down to Bob Quinn Lake is too steep for most vehicles, but the distance is a short easy walk. Coastal western hemlock and white spruce have replaced subalpine fir and mixed birches. As the road swings more northerly, the hemlocks rapidly become more stunted.

**Km 330.0:** Iskut or Burrage Burn; a 1958 fire started by lightning. Hwy 37 passes through about 30 km of "the largest blueberry patch in B.C." Several varieties of huckleberries cover the burn, guaranteeing good picking throughout August. The Canada blueberry is less common, but can be found. Transitional forest is present including pioneer species such as lodgepole pine, saxifrage, and Sitka burnett. Moose, black bear and fox abound. Since tall clumps of willows and alders form a maze among the charred snags, hikers should carry a bear bell.

**Km 361.0:** Eastman Creek has a pullout with tables and toilets. You are through the burn and into an open pine/aspen/spruce forest dotted with black spruce muskegs, similar to that found at Prince George. Birds are interior rather than coastal species.

**Km 365.5**: Willow Creek, Willow Ridge R.V. Park. Steep access; good view of Spectrum Range within Edziza Park in the distance.

**Km 372.0:** Swampy 2 km winter-trail to Natadasleen Lake and Cascade Falls. The Iskut River widens to form Natadasleen Lake. The lake outlet is a dangerous series of ever-increasing cascades, with the river splitting and dropping over a 30 m stepped falls. The best and safest route to the falls is by jet boat. Stop at Tatogga Lake and hire someone to take you through the lakes by jet boat to a well blazed trail. Take this 1 km hike to see Cascade Falls.

**Km 380.0:** Kinaskan Provincial Park campground sits on Kinaskan Lake. The lake is very large and contains huge trout, but you must watch for strong winds.

Todagin Creek. Occasionally goats can be seen on the cliffs above the highway.

**Km 405.0:** Tatogga Lake (summer only). Here you will find a cafe, gas station, lodging and charter boat service with camping along the lake. Tommy Walker, author of the book *Spatsizi*, was the first owner.

**Km 408.0:** Klappan River/Ealue Lake access road. Spruce-pine forest along here has Labrador tea, baneberry and northern bedstraw. Watch for great horned owls; gyrfalcon can be seen occasionally in summer. Moose, black bear and fox are present.

Ealue Lake is the best spot along the Stewart-Cassiar Rd to spend a week in nature study. It has a well-run private campground 12 km off the highway. From here, dozens of old mineral-exploration trails lead easily all over the area, both above and below timberline. Sheep and goats are at least as plentiful as bears. The underbush is less dense than around the Nass & Bell-Irving. You can climb to alpine country in less than an hour.

The road past Ealue Lake continues for approximately 23 km to the Klappan River, at the crossing point for the "trail of 98," now a B.C. Forest Service campsite. Klappan River is the western boundary for Spatsizi Plateau Wilderness Park. Beyond here Gulf Resources is developing a massive anthracite coal deposit. The road is narrow and rough and there may be coal trucks on it.

**Km 413.0:** Eddontenajon Lake, at 12 km. long, is not as long as Kinaskan, but it has the same character: deep water, large trout, strong winds. In spring and fall, migrating swans are seen in the shallows. Some swans overwinter here. Black swift, varied thrush, Oregon junco and Steller's jays that are common south of the Burrage Burn, from here give way to migrating pintail, longspur, horned lark, white-crowned sparrow, slate-colored juncoes. Northern species such as white-winged crossbills and boreal chickadees appear in numbers.

Mt. Edziza is hidden from view by the peaks of the Klastline "Plateau" rising abruptly from the lakes.

**Km 414.6:** Iskutine campground, store, and float plane base on the shore of Eddontenajon Lake. Charter planes are available to tour Mt. Edziza Provincial Park, a vast volcanic wilderness area accessible only by plane or horse. Charter planes are also available out of Telegraph Creek. Iskutine is easily reached by road.

**Km 417.4** Black Sheep Motel: open year-round with gas, food and lodging. It is cold here with dry snow by Halloween.

Across the valley, as you approach Iskut Village you will notice a precipitous mountain with two remarkable features: pinnacles and a rock glacier . The moraine directly above the village has a steep front which shows signs of glacier-like movement and collapse. Behind it lie wave after wave of rubble. Only frost heaving is causing this movement. Little or no ice now remains under the rubble. The pinnacles on the skyline above form the famous "loon's bill," for which the mountain is named.

**Km 420.3** Iskut Village. Every spring horned larks come through here in large flocks. Gas and food are available. This is the take off point to Spatsizi Provincial Park. The village has a grocery store and post office. Ask at the grocery store for directions to the Indian band office, where you can sign on for a regular scheduled trail ride, river-raft trip, or combination of both; or arrange for a guide for the "40 mile trail" or other trip you may plan. Bear Paw Ranch and other accommodation is in the area.

**Km 433.0:** Forty Mile Flats cafe with gas and food available. From just north of here you can take an unmaintained track on the right called "The Microwave Rd" to semi-alpine country dotted with what is called "buckbrush", scrub willows, beautiful yellow-flowered shrubby cinquefoil, bright with wildflowers in season. This road has deteriorated to a 4 x 4/trail bike track.

**Km 439.0:** A pulloff on the left overlooks a black spruce muskeg. The water of Morchuea Lake can be seen among the trees, together with an excellent view of Mt. Edziza. The next side-track to the left leads to disused gravel pits near Morchuea Lake. A popular camping spot although there are no facilities. In good weather, Mt. Edziza sits like a mound of ice cream directly across the lake. Morchuea (Mo choo ee ah) is a proposed park site. Edziza, a huge composite volcano, is covered on this side by an icefield from 2,000 m to its 2787 m summit and locally is called Ice Mountain. It was formed by repeated eruptions of alkaline basalt and pumice. Thirty small lava and cinder cones flank Edziza. Since the last glaciation several small cones have erupted which are not eroded and hence are very recent.

**Km 447.0:** This highway truck pullout gives a good view of the Stikine River and valley, and Gnat Pass to the north.

**Stikine Hill**: Notice the unstable soil. In 1977, miles and miles of this material put a halt to the B.C. Rail construction to Dease Lake, and closed much of the roadbed which had been built.

**Km 452.0:** Stikine River bridge north side is where all water craft, canoes and rafts are taken out of the river. Downstream, the river goes immediately into a sheer-walled gorge, too rough for navigation. This is the beginning of the Grand Canyon of the Stikine. It is 90 km long and at the most dramatic point, visible only by air, 500 metres deep.

The lower Stikine is reached from Telegraph Creek.

**Km 476.0:** Gnat Lakes in Gnat Pass. Here Bohemian waxwing and Arctic ground squirrel are seen all summer. Rock ptarmigan and Arctic tern are found. Caribou and moose may be sighted. Look for mountain spirea, twin flower, shrubby cinquefoil, kinnikinnick, Arctic lupine and spreading phlox.

Gnat Pass summit at 1241 m. As you travel north beyond Gnat Pass note the excellent view of the Snow Peaks to the northwest and Mount McLeod and Dease Lake to the North.

**Km 493.0:** Tanzilla Bridge: A designated campground. Between here and Dease Lake is the low divide (820 m) between the Pacific and the Arctic.

**Km 502.7**: Dease Lake. Gas, restaurant, lodging and a service station for repairs. Limited groceries, R.C.M.P., B.C. Parks, B.C. Forest

Service office, a base for jade mining, jade sales, 2 motels, pub, charter and scheduled air service.

**Telegraph Creek Road**: The road southwest to Telegraph Creek takes off from Hwy 37 in the community of Dease Lake. The intersection lies between the R.C.M.P. building and South Dease Services. Turn left here for Telegraph Creek and most of Dease Lake town. The last 50 km of the Telegraph Creek Rd, the interesting part, has one-lane cliffhanger sections with grades to 20% and very sharp switchbacks. A good trip for cars with good brakes. Trailers or very long vehicles can be problems. This is very dry country, open, and lying in the rain shadow of the Coast Mountains.

About 110 km from Dease Lake, the road leads out onto a point of land. Stop where there is plenty of room on the left, just before the road descends to the Tahltan River. Park and walk ahead to look over both sides. Lava poured into an ancient canyon, disrupting two streams, forcing them to the side of their valleys. The Stikine on the left and the Tahltan on the right began to cut new canyons, leaving a lava peninsula between. It is thought that the rivers took some gravel from under the basalt flow itself, near the point of land which remains today. Then the undermined basalt cracked into huge blocks and settled again onto the gravel. The streams were so entrenched in their new courses that they did not disturb the narrow peninsula of rock again. Today both rivers are a sheer 122 metres below the road.

From river level, note the columnar basalt fanning out on the opposite side. This too was once a river valley filled with lava which formed joints as it cooled. It looks like a huge bird with wings outspread, and is called The Eagle.

Telegraph Creek contains a Tahltan Indian Reserve with a small grocery store. The road then drops sharply into the historic lower town by the river. The community has a choice of outfitters to take you down the Stikine past the Great Glacier all the way to Wrangell, Alaska. It can be canoed by the well-experienced.

The old Hudson's Bay store in Telegraph Creek has been restored as the Riversong Cafe. Sleeping rooms are usually available. Gasoline can be purchased, but there is nothing even remotely resembling a service station in Telegraph. Buildings in the lower town are tall turn-of-the-century wood frame structures facing the river, which was the highway until 1960.

A public campground lies beside Sawmill Lake, above Telegraph and 3.5 km west on the Glenora Rd.

Return to Dease Lake.

Back on Hwy 37, there is a Ministry of Highways Rest Stop campground at Grizzly Creek 35 km north of Dease.

The highway runs parallel to Dease Lake for the entire 45 km of its length. Watch for bald eagle, merganser, mallards, geese and peregrine falcon. Moose, fox and black bear are common. Forests consist of pine trees with Labrador tea, crowberry, twin flower, saxifrage, princess pine, waxberry and meadow cinquefoil.

Dease River, the outlet of Dease Lake, meanders north along a drift-filled valley through the Cassiar Mountains.

**Km 571.0:** Dease River bridge crossing and Joe Irwin Lake. This is a good spot to begin a riverboat or canoe trip down the Dease River. Put in at the crossing (Joe Irwin Lake); go out at Pine Tree Lake, Cotton Lake or at McDame. There are rapids below Pine Tree Lake and Cottonwood River. Use the map NTS Map 104 P. Do not miss McDame. It is only a patch pasture with one log building. There are campgrounds at both ends of Joe Irwin Lake and at Cotton Lake 24 km north.

Through the Cassiar Mountains, Hwy 37 cuts first show banded metamorphic rocks, then by granites of the Cassiar complex batholith, a vast mineralized mountain range with several active mines. Gold, silver, copper, lead, zinc, asbestos, chromite, nickel, and nephrite jade are known. Rockhounds will love exploring the Cassiar, but honour posted areas so you are not panning a placer lease!

**Km 579.0**: Pine Tree Lake on the Dease River is accessible to take boats onto the Dease River system.

**Km 584.0**: The Dease River forest fire of 1982 burned 9,312 ha (23,000 acres). Hwy 37 travels through this burn for eight km. Being in the Arctic watershed, there are grayling in the Dease River and its lakes.

**Km 605.0:** The old road-construction campsite beside Simmons Lake is a good place to stop for some mountain hiking although it is not a designated campground. North of the highway is an old road suitable for hiking or trail bikes. This road crosses Bass Creek on a bridge upstream from the tiny Champagne Lake, and leads through timber, subalpine country, alpine meadows, to an abandoned molybdenum mine on the bare rock skyline, at 2100 metres.

**Km 618.0**: Junction to the Cassiar Mine & Cassiar Townsite. This is a company town where the chief product is asbestos. It lies 14 km west from this junction. Gas and food are available in the town but no camping. Campers go north to Boya Lake or south to Simmons Lake, Cottonwood River, or “Mighty Moe’s” at Cotton Lake.

Cassiar has a grocery and dry goods store plus a cafeteria called the "Cookery."

The Cassiar Mountains are excellent for rockhounds. Quartzrock Creek on the Cassiar Rd has a gold mine. There was a gold rush here in 1874. Stone sheep are seen on the mountains.

**Km 630.0:** Placer gold works can be seen along McDame Creek beside the highway.

**Km 641.0:** Good Hope Lake. The beautiful blue and turquoise colours of Good Hope Lake and adjacent Mud Lake are caused by their shallowness and fine light-coloured silt coming off the glaciers.

**Km 653.0:** Boya Lake Provincial Park is located 3.3 km off Hwy 37, along Boya Lake. Cassiar people call this "Chain Lakes," an appropriate name indicating the complexity of channels and islands which make it a superb boating lake. Because of its shallowness the lake warms up in summer and by mid-July water temperatures reach 18° C. Located on the northeast side of the Cassiar Mountains, northern and prairie species of birds and vegetation are found. The forests are made up of mixed stands of trembling aspen, white and black spruce and pine in a low and rolling landscape. Common terns are seen on the lake. Moose and beaver are common while mountain goat and caribou roam above the nearby timberline. Hawk moths feed in the fireweed at Boya. These giant moths look like hummingbirds hovering. The Horseranch Range, across the lake from the highway and park, is the last thrust of the Cassiar Mountains you will see. They are very old (Precambrian) metamorphic rock, gneisses. A few tamarack trees are seen here and there on the north side of the Cassiar Mountains.

**Km 660.0**: At this point, Hwy 37 enters the Liard Plateau and views of rolling hills remain with you to the junction with the Alaska Hwy. Great horned owl and black-billed magpie are seen. The dry pine forest has reindeer moss, Indian paintbrush, white pussytoes and agoseris.

The Plateau is a flat or rolling plain dotted by black-spruce muskegs, boggy lakes, and gravel eskers left by the melting continental glacier. Streams flow under glacial ice, carrying and sorting sands and gravels which in turn are left as ridges called eskers that follow the meandering pattern of a stream. As the continental ice melted it left an irregular hummocky terrain of unsorted "boulder clay" crossed by these ridges of sand and gravel.

**Km 695.0:** Near Wheeler Lake, a tamarack bog lies next to the highway beside a good pullout.

**Km 735.0:** The B.C.- Yukon Border.

**Km 738.0:** Junction of Hwy 37 and the Alaska Hwy . Gas, food and lodging Watson Lake, lying 24 km east of the junction along the Alaska Hwy, is a substantial community with all services including an airport.

**Further Information**

Several books are worth reading before taking the trip. Norman Lee, *Klondike Cattle Drive,* Mitchell Press, Victoria, 1960, 58pp. Tommy Walker, *Spatsizi,* (an autobiography, about Tweedsmuir Park country for 1st half then Spatsizi). Patterson, *Trail to the Interior.* Allen A Wright, *Prelude to Bonanza,* Gray's Publishing, Sidney, B.C., 1976, 321pp. Edward Hoagland, *Notes from the Century Before: A Journal from British Columbia,* North Point Press, San Francisco, 1982, 273 pp. Ozzie Hutchings, *Stewart: The B.C.- Alaska Border Town That Wouldn't Die,* Stewart, Victoria, 1976, 80pp. *The Stikine River,* Alaska Geographical Society.

# SPATSIZI PLATEAU WILDERNESS PROVINCIAL PARK

Spatsizi, with an area of 674,024 ha, is B.C.'s second largest park. Over 140 species of birds have been recorded including nesting gyrfalcon.

The park sprawls across two broad physiographic regions: the Spatsizi Plateau and Skeena Mountains. The plateau is a rolling upland which extends in a broad curve from Mount Brock in the northwest to Tuaton Lake in the southeast. Valleys are broad and U-shaped. Mountains are the Eaglenest Range, part of the Skeena Mountain system. They are rugged with cirques, tarns, and hanging valleys on their northern and eastern sides.

Bird life is prolific with coastal, interior, northern and southern species seen here.

The Stikine valley on the north side of the park is considered to be critical winter habitat for caribou. The large rainshadow, covered by the plateau, forms one of the most significant ranges for caribou in the province.

Vegetation is divided into three zones, Alpine Tundra, Shrubby Spruce-Willow-Birch, and Boreal White and Black Spruce forest.

**Birds**: Smith's longspur and lesser golden plover, four loons species, three grouse, three ptarmigan, five gull and four owl species, yellow-billed, Pacific and red-throated loons, Canada goose, green-winged teal, and Barrow's goldeneye. Gyrfalcons, Long-tailed jaeger, wheatear and snow bunting, sharp-shinned hawk, northern harrier, golden and bald eagles, ruffed, blue and spruce grouse, willow, rock and white-tailed ptarmigan, woodpeckers, waxwings, warblers and sparrows.
**Mammals**: moose, mountain goat, Stone sheep, caribou (the large "Osborne" variety), grizzly, hoary marmot, Arctic ground squirrel, mule deer, wolves, black bear, wolverine, marten, fisher lynx and beaver.
**Amphibians**: northwestern toad and wood frog.
**Fish**: rainbow trout, lake trout, Dolly Varden, Mountain whitefish, longnose sucker, burbot and Arctic grayling.

**Further Information**

The park is about 300 km north of Smithers and 60 km east of Eddontenajon on Hwy 37. Visitors usually come via aircraft from Eddontenajon or Smithers. Trans-Provincial airlines have float planes at both locations. All major lakes are accessible to these aircraft. It is illegal to land aircraft on Gladys Lake in the Ecological Reserve. Those wishing to ride horses or hike into the park can reach the Klappan River from a secondary road which extends 22 km east from Eddontenajon Lake. The highway communities of Eddontenajon Lake, Tatogga Lake and Iskut each have gas, food and lodging. Cold Fish Lake campsite in the park has five rustic cabins, a sauna and cook house available to the public. These facilities are open all year on a first-come basis with a limit of seven consecutive days stay and no more than 14 days in one year. Primitive camping is permitted throughout the park with several old guide-outfitter camps found in favourable spots. • Primitive trails criss-cross the park. They are maintained by the outfitters as needed. The Parks Division have added some more recently which lead off the B.C. Rail right-of-way. However if you want to get around the park, you generally must find your way by topographical map and compass. You may or may not find a trail to get you where you want to be. • For information on guided trail rides, backpacking, rafting and fishing trips, and photographic safaris that may be offered in the park, contact B.C. Parks and Outdoor Recreation Division, Skeena District, Bag 5000, Smithers, B.C. V0J 2N0. • Topographic sheets include 94E, 104I and 104H. • Suggested reading: Valerius Geist, "*Mountain Sheep and Man in the Northern Wilds,*" 1975, Cornell University Press (a good overview of the sheep and goats of the park); T.A. Walker, *Spatsizi* 1976. Nunaga Publishing of Surrey B.C.

*Material from the Ministry of Environment, Lands and Parks, plus the book* The B.C. Parks Explorer *by Maggie Paquet*

# TATLATUI PROVINCIAL PARK

The park is a remote wilderness area in north-central B.C., on the southern side of Spatsizi Plateau and Provincial Park and on the eastern fringe of the Skeena Mountains. As in Spatsizi to the north, birds abound with 86 species recorded.

Three vegetation zones are present. The Boreal White Spruce and Black Spruce zone is found at the lowest elevations in the Firesteel Valley and extends up Rognaas Creek to the western end of Kitchener Lake.

The Skeena Mountains were raised, folded and faulted about 135 million years ago. The complex folding of the sediments is quite apparent on the slopes at the western end of Kitchener Lake. The overall plateau area is underlain by dipping beds of sandstone, shale, conglomerate and minor coal deposits that are slightly younger than those in the mountains. The mountains and the plateau have been strongly eroded by glaciation. Valleys are deep and U-shaped with signs of ice-scouring along valley walls. Drumlins, moraines and eskers, all made by glacier, are evident at the west end of Trygve and Stalk Lakes in the central and near the north sides of the park.

**Birds**: waterfowl, gray jay, Bohemian waxwing, blackpoll warbler, snow bunting, mew

and Bonaparte's gulls, and rock ptarmigan.
**Mammals**: caribou, moose, Stone sheep, mountain goat, beaver, Arctic ground squirrels, pika, marten, short-tailed weasel, muskrat, grizzly and wolf.
**Fish**: Rainbow trout
**Plants**: Aspen, birch, juniper and saskatoon, Englemann Spruce, Subalpine Fir, Cinquefoil, cow parsnip, lupine, red and white heather, bunchgrass, dwarf birch, alpine willow and Sitka mountain ash.

**Further Information**
The park is located about 240 km north of Smithers and about 180 km southeast of Eddontenajon Lake on Hwy 37. There are no roads into the park. All major lakes are accessible by float plane. Nearest charters for these planes are at Smithers and Eddontenajon where Trans-Provincial Airline maintains a float plane base. Horses and helicopter are other options. Love Brothers and Lee, certified Guide Outfitters, maintain several cabins in the park. They have their own plane for transporting clients into and out of their area. Bob Henderson has a non-hunting guiding service centered on Tatlatui Lake and may be contacted at Tatlatui Wilderness Adventures, RR2, S47 C1, Smithers, V0J 2N0. For further information regarding guiding and other services offered in the park contact B.C. Parks and Outdoor Recreation Division, 1011-4th Ave., Prince George, B.C. V2L 3H9, or the Skeena District, Bag 5000, Smithers, B.C. V0J 2N0. • The Tompkins cabin on Kitchener Lake is open to visitors on a first-come basis. No one can occupy it for more than seven days. • There are primitive trails throughout the park with excellent primitive camping sites on lakes and streams. Often the start of trails are not marked or regularly maintained. • Topographic Sheets include 94D and E.

*Provincial Park information and the book*
The B.C. Parks Explorer *by Maggie Paquet*

# ATLIN PROVINCIAL PARK

Fully one third of this wilderness area is occupied by glaciers, the most prominent being Llewellyn. This large mass of ice has a tongue which sweeps down from the Alaska border nearly into Atlin Lake. The lake, which is coloured a brilliant aquamarine hue from the meltwaters off the ice, lies in the headwaters of the Yukon River system. Atlin is reported the second largest natural lake in the province. Theresa Island, located in the southern part of Atlin Lake is reported to be the worlds largest inland island.

Vegetation in the park consists of boreal forest, subalpine and alpine tundra.

Sedimentary and volcanic rocks of Palaeozoic and Mesozoic eras make up the principal rock types in the park. The entire area was covered with ice during the Pleistocene. This is one of the few areas in the province to see glaciers still at work.

**Birds**: Arctic terns and three species of ptarmigan.
**Mammals**: Stone sheep, goats and Osborne caribou, black-tailed deer, moose, black bear, grizzlies, foxes, hoary marmots, Arctic ground squirrels, pika, beaver, otter and wolf.
**Fish**: lake trout, two species of whitefish and Dolly Varden char. Bonaparte's gulls, Arctic

terns, common loons, blue grouse, rock, willow and white-tailed ptarmigan.
**Low-elevation Plants:** white spruce, lodgepole pine, trembling aspen, white birch and cottonwood.
**Mid-elevation Plants:** subalpine fir and Engelmann spruce.
**Alpine tundra**: heather, dwarf willow, juniper, mosses, lichen and some grasses.

**Further Information**

Reach Atlin by turning south at Jake's Corner on the Alaska Hwy . Drive 100 km south on an all weather gravel road to the town. Access into the park is by aircraft and/or boat charter. Local guides are available for natural history exploration, fishing or hunting. • There are no facilities within the park. The town of Atlin offers a complete range of modern amenities. Wilderness camping is permitted throughout the park. Take standard precautions for bears both in camp and on the trails. • There are no roads or maintained trails in the park. Overgrown trails lead from the head of Skoko Inlet to Skoko Lake and to Llewellyn Glacier. A short trail leads from the head of Llewellyn Inlet to a viewpoint overlooking the glacier. Foot crossing of the O'Donnel River is impossible except at low water. Access to alpine hiking is available from the south-east slopes of Birch Mountain and Mt. McCallum. • Boaters in small craft should be very cautious on this lake as these waters are subject to sudden, strong winds originating from the Coast Mountains.

*Notes from B.C. Parks*

# HAINES ROAD

The Chilkat Pass area is one of the best places to find many species of birds that usually nest in the Arctic.

Vegetation is exceedingly varied with several ecotypes, including northern boreal forest, subalpine forest and scrub, alpine tundra and coastal rain forest. The subalpine and alpine areas provide the best floral show in July.

Geological features include the Nadahinni Ice Caves, a cave at the end of a glacier. To find the cave go upstream on Nadahinni Creek for about five km. A rock glacier is found near the side of the road just north of Dezadeash in the Yukon. Look for the sign on the side of the road.

**Birds**: common redpolls, rosy finches, red-necked phalarope, gyrfalcons, eagles, willow (most common), white-tailed and rock ptarmigan, Smith's longspur, gray-crowned rosy finch, and red-necked phalarope, Peregrine falcons, blue grouse, sea ducks, Baird's and least sandpipers, red-necked phalarope and short-billed dowitcher, mew gull, Arctic tern, spruce grouse, boreal chickadee, goshawk, boreal owl, great horned owl, several thrushes including gray-cheeked, fox sparrow, snow bunting, and northern shrike.
**Mammals**: grizzly, black bear, wolf, coyote, red fox, moose, wolverine, Arctic ground squirrel, river otter, collared pika, and various voles.

**Further Information**

Haines Rd extends for 220 km from Haines Junction on the Alaska Hwy in the southern Yukon, south through the northwest corner of B.C. and down to Haines, Alaska. There are three ways to access the road: take an inland passage ferry from Prince Rupert, B.C. to Haines, Alaska, and drive north; fly to Whitehorse, rent a vehicle and drive south to Haines; or drive up the Alaska Hwy and continue south past Kluane National Park to Haines. • There are four campsites in the area: Pine Lake which is 10 km north of Haines Junction, one at Kathleen Lake, another at Dezadeash Lake in the Yukon and on the Haines Rd, and the fourth at Million Dollar Falls also on the Haines Rd. Motels include five at Haines Junction, one at Kathleen Lake (Kathleen Lake Lodge) and several in Haines. Gas stations and garages are in short supply. There are several in Haines Junction, and no more until Haines. • Topographic map is 114 P, Tatshenshini River. The article "Birds of Chilkat Pass British Columbia" by R.B. Weeden, 1960. Can. Field Nat. 74: 119-129 will supply some help. The Kluane National Park Interpretation Centre in Haines Junction has check lists, maps and information. Weather can be very variable so come prepared for wide temperature fluctuations.

*Susan Hannon*

# INDEX TO SITES AND PLACES

## H



## I • J



## K



## L



## M

## N



## O



## P

## T



## U



## V



## W



## Y

# ABOUT THE EDITORS

With a life-long love of nature, Joy and Cam Finlay have contributed, with enthusiasm and knowledge, to many natural history projects since the mid 1960's. Joy led some of the first regularly scheduled urban nature walks, taking families and school classes in the out-of-doors in Edmonton. That same year Cam was hired as one of the first permanent park naturalist interpreters with the National Park Service.

Joy's work expanded to include giving programmes, courses and sessions across the continent. After initiating snowshoeing programmes she wrote and published *Winter Here and Now*, a best seller written for leaders, teachers and parents. In addition she has published numerous articles which illustrate activities and encourage people to explore in the out-of-doors.

After developing a major outdoor historical museum, Fort Edmonton Park, Cam became the founding director of the John Janzen Nature Centre, the first major urban nature centre to open in Canada. He was also the compiler and editor of *A Bird-Finding Guide to Canada*.

They were the driving force behind the compilation and writing of the book *A Nature Guide to Alberta*. Then, using a large collection of reports on municipal and provincial parks in Alberta, they wrote *Parks in Alberta — A guide to peaks, ponds, parklands and prairies for visitors*. They also wrote other booklets including *Feeding Birds in Alberta* and *Who Goes There — Tracks and Signs of Wildlife*.

They write a weekly column on birds and other aspects of natural history for the *Edmonton Journal* and have continued this project for the past seven years. They wrote a similar column simultaneously for the *Calgary Herald* for three years.

Both members of this wife/husband team have been recognized with numerous awards at local, provincial and national levels for their work in conservation.

# Other Books From Lone Pine Publishing

***Discoverer's Guide: Fraser River Delta* — Don Watmough**

A recreational and nature guide to over forty sites along this ecological and economic lifeblood of southwestern British Columbia, from Boundary Bay to Pitt River. Full-colour photos and maps.

$12.95 softcover   128pp.   5 1/2 x 8 1/2   ISBN 1-55105-014-5

***Plants of Northern British Columbia* — MacKinnon, Pojar, Coupé**

A handy reference guide to common plants of the northern interior region of B.C. for the amateur or professional botanist, naturalist, or forester. Includes keys, conspectuses, and detailed notes on ethnobotanical and etymological information for each species. Full-colour photos and line drawings.

$19.95 softcover   352 pp.   5 1/2 x 8 1/2   ISBN 1-55105-015-3

***Bicycling Vancouver* — Volker Bodegom**

A must for all Vancouver and area cyclists, this touring guidebook includes over 30 routes in and around Vancouver from Horseshoe Bay east to New Westminster and Delta. Each route includes a map and a detailed road log. Information on clothing, equipment, safety, different types of cycling, advocacy, and clubs is included in the extensive reference section. Maps, photographs.

$14.95 softcover   208 pp.   5 1/2 x 8 1/2   ISBN 1-55105-012-9

***Gardening in Vancouver* — Judy Newton**

Tailored to Vancouver's soil and climate conditions, and intended for both intermediate and home gardeners. Contains two full-colour sections with expert hints for successful gardening. Full-colour photographs, illustrations.

$14.95 softcover   176 pp.   5 1/2 x 8 1/2   ISBN 0-919433-74-X

***Mushrooms of Western Canada* — Helene M.E. Schalkwijk-Barendsen**

The first comprehensive field guide to the mushrooms of western Canada and the Pacific Northwest. Describes 550 species, including information on habitat, distribution, identifying characteristics and edibility. Full-colour illustrations.

$19.95 softcover   416 pages   5 1/2 x 8 1/2   ISBN 0-919433-47-2

***Picnic Guide to British Columbia* — Nancy Gibson & John Whittaker**

An entertaining and unusual guide to picknicking, including details of local history and providing suggestions on things to see, to do, and (most importantly) to eat. Forty sites, maps, and black and white photos.

$11.95   264 pp.   5 1/2 x 8 1/2   ISBN 0-919433-59-6

Buy these and other books from your local bookstore or order directly from Lone Pine Publishing, #206, 10426 - 81 Avenue, Edmonton, Alberta, T6E 1X5. Phone : (403) 433-9333  Fax: (403) 433-9646